DALLAS EXPOSITION.
THE PEOPLE'S GREAT ANNUAL CARN

The Great State Fair of Texas

$1 2.00

C50

The Great State Fair of Texas

AN ILLUSTRATED HISTORY

Nancy Wiley

FOREWORD BY
A.C. GREENE

TAYLOR PUBLISHING COMPANY
Dallas, Texas

This book is dedicated to the memory of
Robert Brooks Cullum
President of the State Fair of Texas
1966-1978

1550 West Mockingbird Lane, Dallas, Texas 75235

Library of Congress Cataloging-in-Publication Data

Wiley, Nancy (Nancy N.)
The great State Fair of Texas.

Bibliography: p.
Includes index.
1. State Fair of Texas — History. 2. State Fair of Texas — Pictorial works. I. Title.
S555.T42D359 1985 976.4 85-17391
ISBN 0-87833-465-3

Book Design by Bonnie Baumann

Printed in the United States of America

CONTENTS

FOREWORD

There is scarcely a native of North, East, Central or West Texas, particularly those of us with silver threads among our gold, brunette and red, who does not have separate and personal memories of the State Fair of Texas. Though I was neither a Dallas native nor nearby resident, my own recollections go back to the mid-1930s when I stood with great pride and youthful anonymity on the steps of the Hall of State to watch my great-grandmother Longley honored as one of the last surviving widows of the Texas Revolution (she lived to be the last). I can remember attending a college football game in the Cotton Bowl before it was called the Cotton Bowl. And I can recall riding a streetcar to the front gate in the days when all trolley cars were routed via Fair Park during the show — never realizing, at age 9 . . . 10 . . . 12, that one little girl who might have gotten aboard along Cole Avenue would someday be my wife.

My memories skip forward to those sparkling autumn days of the Fifties, listening to Big Tex for the first time, riding the creaky aluminum monorail as it made its loop above the crowds. Then in the Sixties, when my family and I moved to Dallas, watching a batch of children (my own) grow up at the fairs, gradually graduating from the gentle, colorful carousel to the fierceness of the Wild Mouse and, ultimately, the Comet Roller Coaster; dragging me (hurry, Dad!) to the Midway, the Age of Steam, the Dr Pepper Circus — and always, always there was the food: first (and forever) a Fletchers' Corny Dog and a hamburger from one of the many booths, or one of those ice cream bars rolled in chopped nuts — all of which still taste best when eaten at the State Fair. Then came the brilliant Belgian Waffles and a short time later, buttered corn on the cob . . . all of us sharing sweltering days and (now and again) cool nights . . . and walking to the car, each of us, mother and dad, holding one asleep on a shoulder and another by the hand.

Personal and private memories, but not unique, because my recollections of the State Fair can be duplicated in the memories of hundreds of thousands of Texans. And different, but similar remembrances can be found among those who came to the State Fair from all the little towns who used to keep individual identity. Some have tender recollections of earlier days, riding the electric interurbans in from Waco, Denison, Corsicana, and all points in-between; the excitement as the bouncy local cars swung over the transit trestle from Oak Cliff and eased down to the little station at Ferris Plaza to pick up the fairgoers who had arrived at Union Station; then later, tired but happy, zinging back home through the dark, the sparks flying upward from the overhead wire. And there are still those, getting scarce now but with memories just as strong, who can recall riding the steam cars or loading up the wagon to go to Dallas and "the Fair!" There has been magic in the phrase and the thought for a century.

It is pure sentimentality, of course, to remember only perfection. That's the way we tend to think about happy times. The mosquito bites, the sprained ankles and aching feet, the incessant demands of one child or another, to be fed, or watered or de-watered . . . the persistent pleas to ride something they were unfit, as yet, to ride, or play some game they were unable to perform — these annoyances and inconveniences have disappeared with time. Nostalgia is particularly appropriate for something as ever-rolling and never-ceasing as the State Fair of Texas. It is, one of those aforementioned weary children, now grown, asserts, "the last great Dallas tradition."

Yet, through all the years of recollection, of romanticism and sentimentality, no one wrote the story of the State Fair of Texas. Astounding, isn't it?

The Great State Fair of Texas is an *inside* story, not a fairgoer's remembrance. Readers will find it an enlightening, valuable book, because it tells, in great detail, the *why* of the things most of the fairgoing public has taken for granted: that the exposition would, indeed, be open each fall, that the familiar would be spiced with the new, that such stable items (no pun, please) as the horse and cattle barns, the hog pens or the skirt-lifting blast of air at the fun house would be there, that the most important thing in the world might be to win one of those ridiculous plush animals.

The intricate thread of development which not only changed but on occasion eliminated some of these constants (the horse-racing set-up, for one major instance) is given clear and authentic narrative in Nancy Wiley's history. To anyone who has grown up with the State Fair of Texas, whether that growing up consisted of long-ago visits or dates from a few seasons ago when you were first allowed to go to the fair with "the gang" rather than the family, or on that never-to-be-forgotten first date, this story is amusing and intriguing. To the historian, it is full of treasure.

And behind all the facts, the history, the changes and growth, lies that one basis for both recollection and continuation: enjoyment. As John T. Trezevant, president of the Fair Association in 1895, sagely observed:

"Experience has taught management that visitors to the State Fair — the majority of them — attend more to seek amusement and be entertained than as a matter of education, and while we propose to preserve the standards of the various departments . . . we will, in a particular manner, look after the attraction programs."

And now, nearly a century later, entertainment is why there is still a booming State Fair of Texas taking place every October. And while Dallas and the Fair Association may groan occasionally at the enormous labor and strife needed, from time to time, to put or keep things aright — entertainment is why the fair will continue into the 21st century and probably well beyond. The nature of the exhibits will certainly change — livestock and agricultural exhibits were still emphasized long past the time Texas, particularly the northern area, was agrarian — and even the nature of the thrill rides will advance with the years; but as John Trezevant said, ". . . the majority attend more to seek amusement and be entertained than as a matter of education." American society, not just the Great State of Texas, demands to be entertained first, then it will accept small doses of instruction and enlightenment. The State Fair of Texas is fun. None of the upstanding men who have led the fair through its first century have been able (or inclined) to dramatically change that.

But if this essay seems to make the story of the fair frivolous, then I do a disservice to this very readable and valuable book. True, it is a history; but also true, it is entertaining, as good histories and good fairs invariably are. The history of the State Fair (to use its simplest, longest-lasting title) has paralleled the history of Dallas and reflects community views rather remarkably. While there have been scenes of upset and pressure behind the fair's front gate, none of these scenes extended into the public arena. That alone may mark the fair's greatest success and its ultimate reason for longevity.

Is there a more significantly *Dallas* institution than the State Fair of Texas? I doubt it. What other agency has drawn the best of the city's business and commercial leaders, without pay and presumably without profit, year after year? What other attraction has succeeded, never drastically changing its appeal, in drawing over sixty million patrons through its gates, even in a hundred years? What other organization has contributed such symbols and celebrations to Dallas' fame as Big Tex, the Texas-OU game and the Cotton Bowl Classic? And what other portion of Dallas has remained such a landmark of geographical and spiritual reference as Fair Park?

This book is also a history of the grounds on which all these many fairs and expositions have taken place, and in this respect can be as interesting to the non-fairgoer as to those "never miss a year" persons. The site of Fair Park (or "the fairgrounds") has not moved since 1886, and this location influenced Dallas' development and growth as a city. For instance, in the latter years of the 1880s, a steam railroad was built from downtown out of what was then Forest Avenue (now Martin Luther King Jr. Boulevard) to deliver visitors to the park. Having such a convenient and rapid means of hourly conveyance to the grounds meant the Fair Association had to find something to occupy visitors during the months when the fair itself was not taking place — which it did, with varying degrees of success.

Several associative satellites, we might call them, sprang up adjacent to the fair grounds; William Gaston, who should be credited as father of Fair Park, if not of the fair itself, developed Gaston Park about where the Music Hall stands today, and the Dallas professional baseball team played there, as well as at another northeasterly remove for a period. Famed "Cycle Park" was also located on land incorporated in the modern grounds, and Dallas' first golf course, six holes, or rather peach cans, was sunk into territory along the northern side in 1896, a few months before a "real" links was set up along Turtle Creek in Oak Lawn.

The steam railroad, along with several other street railway lines, was electrified in 1891, and several thousand fair visitors crowded the cars for the trip from the various railroad stations (there were six at the time!) — making the electric trolley cars as much a part of that year's program as the horse races. Those races, by the way, were often dominated by Colonel Henry Exall's horses, bred at his Lomo Alto horse farm which in 1907 was made into a fashionable subdivision north of Dallas: Highland Park.

Regardless of its many years of successful presentation, early days with the State Fair were anything but placid, easy, or harmonious. Rivalries, mishaps, fires, lawsuits and financial woes plagued the organization for many years.

On the whole, however, the fair's grand moments have far outnumbered its defeats. Look at the collection of names that have appeared at the State Fair: Booker T. Washington, Buffalo Bill and Annie Oakley (they drew a record crowd of 70,000), President William Howard Taft, Enrico Caruso, Chief Quanah Parker, Harry Houdini, Hoot Gibson, Jimmy Doolittle, Admiral Chester Nimitz, Audie Murphy, and King Olaf of Norway.

Consider such top attractions as the new motion picture machines featured in 1897 and the first auto races (two cars) staged in 1901. That same year, the fair opened its first Mexican restaurant and introduced the first incubators, or "wooden hens," as they were called.

But despite the organization's enviable record for innovation, State Fair leaders were slow in responding to the times, in the late 1950s and '60s. Successful integration came as something of a shock and proved groundless their fears that white fairgoers would never ride the same rides with blacks, or mingle on the same fairgrounds day in and day out.

So . . . as the State Fair of Texas enters its second century, with suitable predictions of greater glory, its first 100 years are offered to us in delightful and candid detail in this book. It is compiled and written by someone who not only has done a painstaking job of gathering information, but has kept the human side of our Great State Fair foremost in her gathering. All of us — Dallas natives, nostalgia buffs, those silvery-haired rememberers who reach back through many a calendar, and the restless youngster who "can hardly wait" to return to the Midway — can raise a chorus of thanks that she has performed thusly.

A.C. Greene
Dallas, Texas, U.S.A.

AUTHOR'S NOTE

"OUR STATE FAIR IS A GREAT STATE FAIR," wrote lyricist Oscar Hammerstein II, and composer Richard Rodgers added a bouncy tune. No matter that the collaborators were New Yorkers, or that they were thinking about the Iowa State Fair when they wrote the song, Texans are convinced that the musical sentiment describes the State Fair of Texas.

Which, of course, it does.

This book emphasizes the phrase: "GREAT STATE FAIR," but importantly it is also "OUR STATE FAIR" — a showcase of Texas chauvinism and pride for the thousands who take part in each annual exposition.

So too, then, is this "our" State Fair book. The author was encouraged by Taylor Publishing Company to tackle the research and writing, but the story belongs to those individuals — past and present, inside and outside the organization — who recorded, reported, recollected and made history.

The book could not have been completed apart from the generous help of Marjorie Dawson, Jim Skinner, Don Clark, Frances Lewis, Elizabeth Peabody, Claude Perry, Dick Potticary, Claude Wreyford, and the rest of my co-workers at the State Fair of Texas; Peggy Riddle, Sarah Hunter and Beth Bucy of the Dallas Historical Society; Bill and Jean Carpenter of Carpenters & Associates; Shirley Oxendine of SETO Communications; Harry Ellis of Dr Pepper Company; Brad Bailey and Clint Grant of the *Dallas Morning News;* Andy Hanson of the *Dallas Times Herald;* photo archivist Travis Dudley and the entire staff of the Texas/Dallas History and Archives Division of the Dallas Public Library.

Special thanks are due Jim Black for initiating the project; Joseph B. Rucker, Jr., Tom Hughes and Joe M. Dealey for insights and direction offered in extended interviews; David Nixon, Jim Bradley and Charles Kavanaugh for assistance in locating and copying rare photographs and collateral materials; Gainor Eisenlohr for her capable research on the postwar fair; and Bob Halford for the daily support that made this work possible.

I am deeply grateful to Wayne Gallagher, Jeanne Baker, Peggy Riddle and Russell Smith for reading selected chapters for historical accuracy; Barbara Sampson, Cliff Hopper and A. C. Greene for commentary on the entire manuscript; designer Bonnie Baumann; art director Kathy Ferguson and my editor, Freddie Goff, of Taylor Publishing Company for their significant contributions to this book.

On a personal note, I would like to express appreciation to my mother, Minta Newell, for her unwavering encouragement, and to my brother, Doug, for his interest and gratis legal advice. Thanks also to the next generation of Wileys: Robyn, Mike and Lisa, for many months of sustained enthusiasm, and Steve, for setting up the word processing system that permitted me to finish this project on time.

Finally, I am indebted to Jeanne Baker, Bob Halford and Cliff Hopper, whose views and analysis of the fair, shared informally over the past decade, at work and after hours, are reflected in these pages.

Nancy Wiley
August 1985

Opening Days

1859-1886

Bright sunshine washed over the front steps of the three-story Merchants Exchange. Flags fluttered from poles anchored in the pressed brick facade; freshly-cut evergreen branches covered the window sills.

Patriotic bunting swung in giant loops from the cornices of buildings on both sides of Commerce Street including the opulent Grand-Windsor Hotel. The Windsor, which advertised luxurious accommodations at rates of $3.00 per night, was filled to capacity.

Looking east toward the Merchants Exchange Building at the corner of Lamar and Commerce streets where the Dallas State Fair parade formed on the morning of October 26, 1886.

Shortly before nine on this crisp October morning, clusters of fashionably attired guests emerged from the hotel and strolled leisurely along the planked sidewalk, past the printing plant and offices of the one-year-old *Dallas Morning News,* past City Hall to the corner of Lamar where a crowd was gathering.

In the center of the dusty street, parade marshals, wearing shiny badges of authority, shouted instructions while they attempted to organize several hundred men and horses into a formal procession.

The police and sheriff's units, astride carefully-groomed, disciplined mounts, waited attentively. Meine Brothers Band had been assigned the second position, followed by the Dallas Lightguards and Artillery. The Busch Zouave, a military drill squad sponsored by the St. Louis brewery, lined up just ahead of the celebrated Mexican National Band. After the musicians, an open carriage carried former Governor Oran "Old Alcalde" Roberts and his staff. Other dignitaries — Dallas Mayor John Henry Brown, Mayor George Crutcher of East Dallas and various exposition officials — took places behind the governor's party. On down the street, a local bicycle club balanced on high-wheeled roadsters and watched for the signal. Festive banners had been draped over city fire trucks and steamers to establish a rear guard.

As the tower clock in the granite courthouse struck nine, the mounted patrols urged their horses forward and made a right turn onto Lamar. The Dallas State Fair and Exposition of 1886 was finally underway.

The Fifth Dallas County Courthouse, constructed in 1881 and destroyed by fire in 1890.

Spectators stood two and three deep, cheering the parade as it briskly marched and rolled along its one-mile course. For the fair's officers, the route, north and east through the commercial district, prompted glowing smiles. The caravan passed Bartholomew Blankenship's new warehouse, Alex Sanger's expanded dry goods store, the wholesale grocery partnership of John Armstrong and Tom Marsalis, James B. Simpson's law offices and numerous properties financed through the banking interests of E. M. Reardon and W. H. Gaston.

Whatever marks these men had made on Dallas and Texas history, their most lasting contribution would be the one they were now celebrating, though there was scant reason at this time for such a prediction. In eight months, this ambitious project had cost its directors hours of physical labor, additional hours of acrimonious debate, the friendship and goodwill of fellow business leaders, plus an up-front expenditure exceeding $179,000, most of which had been borrowed on the directors' personal notes.

The parade concluded at Union Depot, a massive station constructed the previous year at the point where the north-south line of the Houston & Texas Central Railroad crossed the east-west tracks of the Texas & Pacific. Special guests, band members and officials were transferred to waiting T&P cars for a journey of less than two miles to the new fairgrounds.

The civic enterprise and behind-the-scenes machinations that influenced the railroads to locate their lines through Dallas in the early 1870s sparked a period of tremendous growth for this small village on the banks of the Trinity River. Important mercantile and manufacturing businesses followed, and the population jumped from around 2,000 to 10,385 at the end of the decade.

By 1886, that figure had tripled. Telephone and electric service were available. Major streets had been paved with bois d'arc blocks, a process soon replaced with the development of a rolled surface of crushed stone and gravel called macadam. Hundreds of new commercial buildings, constructed in the prevailing Victorian Italianate style, altered the appearance and boundaries of downtown Dallas.

Union Depot was built by the Houston and Texas Central Railroad on land once used for the earliest Dallas County fairs.

Corpus Christi lays claim to the first fair in Texas. The ambitious two-week event in 1852 was a thinly-disguised land development scheme with murky ties to a revolution brewing in Northern Mexico. Perhaps because the public sensed the ruse, the event was not totally successful — in fact, one historian described it as a "flop."

In 1859, on the same land where Union Depot would be built, the first Dallas fair on record attracted more than 2,000 visitors to a four-day festival that was organized around the area's agrarian interests. Farmers and ranchers from all parts of Dallas County gathered for livestock sales, exchanged ideas and inspected the newest farm implements. A second exposition in 1860 proved even more popular, drawing 10,700 over five days, but fairs were soon forgotten as Texans marched off to defend the Confederacy. Postwar efforts by the Dallas County Agricultural and Mechanical Association to revive the annual event in 1868 and 1869 met with little success.

The North Texas Agricultural, Mechanical and Blood Stock Association, incorporated in 1870 with W. H. Gaston as one of its officers, voted to hold a fair the following year. This plan was aborted when the city donated the original fairgrounds to the railroads for a depot site. The association selected a new location in East Dallas, but preparations for an 1872 exposition were hurried and disorganized, resulting in a limited exhibit program and a stock show that boasted only one registered bull. Significant improvements were made for 1873, even as the nationwide financial panic of that year swept across most of North Texas. Although the Dallas economy remained relatively stable, neighboring communities were severely affected, and the fair was a failure.

No further attempts were made until 1876, when merchants, enjoying the multiple benefits that accompanied the city's emergence as a rail transportation center, concluded the time was right to promote local industrial interests. A fair was scheduled that permitted Dallas to join St. Louis and Kansas City on the major exhibitors' circuit. Stables, a racetrack, and three exhibit halls were constructed, and 30,000 fairgoers swarmed over the grounds during the six-day run. The 1877 event attracted the attention of businessmen throughout the country. Having achieved their goal, Dallas leaders then disbanded the fair.

1886

Progress and prosperity were on display the morning of October 26, 1886, as the train carrying the governor and other guests pulled into the new fairgrounds depot and discharged its passengers.

The location selected for the first Dallas State Fair and Exposition was a table-shaped, 80-acre tract with natural watersheds on three sides and the Texas & Pacific cut as its northern boundary. Situated less than a half-mile from the 1872 site in an area criss-crossed by four rail lines and served by country roads from Rockwall and Kaufman, the fairgrounds were readily accessible via Exposition Avenue, a broad street built that summer as a connecting link to the eastern edge of downtown, or by mule-drawn streetcar along an extended line that ran two miles from the County Courthouse.

A streetcar line provided transportation to the fairgrounds' main entry gate.

A thick, eight-foot-high board fence enclosed the grounds with a circular sweep at the main entrance. Fairgoers, whether on foot or in horse-drawn vehicles, were admitted through an attractive wooden pavilion. The charge was 50 cents per person or $1.00 per vehicle. Parking, in the form of buggy space and horse hitches, was 50 cents extra.

Throughout the summer, local residents had observed increasingly frenzied efforts to meet construction deadlines, but neither they nor out-of-town visitors were fully prepared for the enchanting vista that greeted early arrivals on opening day. In less than six months, a herculean labor of man-and mule-power fueled by large sums of money had transformed acres of bald prairie into a first class exposition park.

The amenities included gardens, five miles of graveled walks and drives, an independent water system supplied from wells and pumped by windmills into a connecting line of troughs and tanks, and a small electric light plant. During the final week of preparation, 200 mature cedar trees were planted along the driveways, and two carloads of cactus were brought in from the Sierra Blanca Mountains.

The centerpiece of the site was a one-mile racetrack built for $10,000 under the supervision of Judge J. H. Dills of Sherman, reputed to be the best track man in North Texas. Fearing that the black waxy dirt in this area would be unsuitable for a fine racing surface, Dills directed the purchase of a branch bottom farm along White Rock Creek, and sandy loam topsoil was hauled in to plate the track.

Approaching the racetrack from the Exposition Avenue entrance, visitors passed the Main Exposition Hall on their left and the Machinery Hall on their right before bearing further

An artist proposed this layout for the Dallas State Fairgrounds in 1886, but the structure to the right of the Exposition Hall was never built.

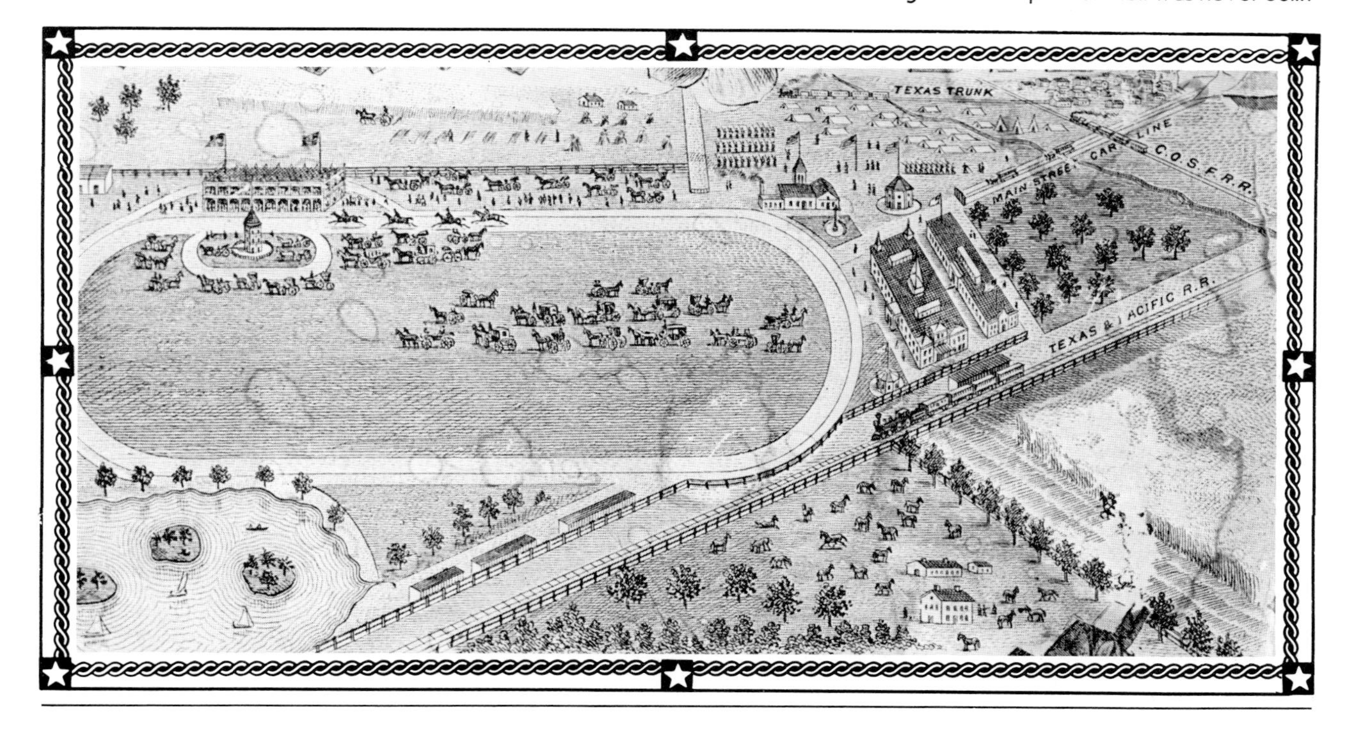

Eight days of horse racing were scheduled at the Dallas State Fair with an $11,000 purse plus $30,000 in livestock premiums.

right, beyond the amusement stands and restaurants, to reach a two-story, 8,000-seat grandstand which had been carefully located to avoid late afternoon glare in spectators' eyes. Directly across the track on the infield was the judges' stand, and stables for 500 horses had been built along the fence. Some distance past the stables, a pretty grove provided an encampment area for military groups, and around the curve of the track at the back of the fairgrounds stood 700 stalls for cattle and other livestock.

By far the most imposing building on the grounds was the Main Exposition Hall. Designed by James Flanders, a leading Dallas architect, the mammoth structure measured 200′ x 300′, roughly the size of a football playing field plus sidelines. Sweeping 75′ arches supported a dome over the building's two-story core, and 100′ towers anchored each of the four corners. The interior was compartmentalized to house the geology, horticulture, education and music halls, as well as the art galleries and ladies' department.

Nearby, the Machinery Hall, a single level, 75′ x 300′ structure also designed by Flanders, featured an outdoor exhibit area devoted to traction engines, hay presses and saw mill equipment.

Opening day ceremonies attracted an enthusiastic crowd to the plaza in front of the Exposition Building. After a spirited overture by the 56-member Mexican Band, Mayor John Henry Brown welcomed his fellow citizens, reminding them that this part of Texas was only a few years removed from "a howling wilderness, roamed over by wild Indians and wild animals." He pointed out that Dallas now boasted 36 churches, 20 schoolhouses and 68 factories. What he did not mention was that, as of this moment, Dallas also had two state fairs; but the existence of a smaller crosstown-rival fair had not escaped anyone, least of all former Governor Roberts who was opening his second fair in this city in as many days. Being a practical politician, Roberts simply recycled his speech from the preceding afternoon. It was mercifully brief. As the

The program cover for 1886 showing the new Exposition Hall.

formalities concluded, the audience streamed through the Exposition Hall doors for a first look at the promised wonders.

Among those thousands enjoying the music, pageantry, sunshine and crowd excitement, perhaps no one individual felt a greater surge of pride than William Henry Gaston.

Alabama-born Billy Gaston grew up on an East Texas farm. Like most of his contemporaries, Gaston's transition to manhood was hastened by the Civil War. With his brothers, Robert and George, he enlisted in the Confederate Army in the spring of 1861.

Neither brother would survive the conflict, but war thrust young Billy into a leadership role five months after his enlistment. Though he was still a few days shy of his twenty-first birthday, the men of Company H of the First Texas Infantry Regiment elected Gaston as their commanding officer, earning him lifelong recognition as the "Boy Captain" of Hood's Texas Brigade.

Captain Gaston led troops into bloody battles in Maryland and Virginia. Recovering from typhoid fever, he was sent back to Texas to recruit more men. Before he could rejoin his company, Union forces took control of the Mississippi, effectively splitting the Confederacy. Gaston served out the war as a purchasing agent.

After discharge, he returned to the family farm in Anderson County. Three years later, following the death of his first wife and his subsequent remarriage to her sister, Gaston packed $20,000 in gold in his saddlebags, and the newlyweds set out for Dallas on horseback.

Dallas was a small frontier town with a population of less than 2,000. Together with an Anderson County friend, Aaron Camp, Gaston purchased a lot across from the courthouse and formed a partnership to open the city's first permanent banking operation. The Gaston and Camp Bank prospered and eventually concentrated its investments in crops, livestock and real estate.

Billy Gaston, with partners and independently, continued to expand his business and personal holdings. Most significantly, in 1871, he purchased 400 acres just outside the eastern boundary of Dallas, property that ten years later would become the heart of a separately incorporated and geographically larger city — East Dallas.

Opportunity followed opportunity, and Gaston, within five years of his arrival in Dallas, had demonstrated remarkable business acumen tempered with a sense of community responsibility. His involvement extended from the organization of local fairs to such major projects as a new iron bridge over the Trinity River and the city's first streetcar line. And by donating land for the required right-of-ways, he provided the key in successful efforts to attract railroad lines to Dallas.

Gaston was a tall, thin, good looking man who loved hunting, fishing and horse racing. Soft-spoken and unassuming, he rejected election as councilman and mayor in East Dallas, preferring to participate in community affairs without the trappings of public office.

James B. Simpson was elected as the first president of the Dallas State Fair and Exposition in 1886.

Billy Gaston's experience with the early Dallas county fairs of the 1870s led him to join other business leaders in preliminary planning for an 1886 exposition of unprecedented magnitude and scope. A charter was granted on January 30. Prominent lawyer James B. Simpson was elected president, and Gaston was selected as one of nine directors for the proposed Dallas State Fair and Exposition.

The old fairgrounds were available to the fledgling organization, but the group decided to purchase new property. A committee was formed to examine all options, and several individuals came forward to suggest tracts of 50 or more acres under varying conditions of sale.

Encouraged by associates, Gaston acquired three pieces of property from his East Dallas neighbors at a cost of $16,000 and offered to sell 80 acres to the association for $14,000, payable in stock.

Members of the directory representing the farm implement, machinery and vehicle interests in the city strongly opposed this selection. Criticism centered on the quality of the rich black soil, which was deemed too waxy for exhibiting heavy equipment. C. A. Keating assailed the property as "the worst kind of hog wallow."

An alternative site of 100 acres on the Cole farm about one mile north of Union Depot was proposed. But when the appeals committee rejected these arguments and reaffirmed the earlier decision, the association split in two. General manager Frank Holland resigned and together with former directors Jules Schneider, O. P. Bowser and Keating, secured a charter for a separate event — the Texas State Fair and Exposition, which they announced would open the day before the Dallas State Fair.

The pros and cons of two state fairs were discussed in the newspaper, on street corners and probably during meetings held at City National Bank, since leaders of both factions served together on the bank's board. Livestock producers objected to the necessity of choosing sides in the dispute, and local merchants groused that they would be forced to assemble duplicate exhibits to avoid offending either group.

Bitterness increased to the point that W. L. Wright of the Moline Plow Company filed suit against the Dallas State Fair accusing Simpson, Gaston and the others of having formed a syndicate to assure selection of a site that would benefit property they already owned.

In late spring, men, machines and mule teams arrived at both locations, and both associations announced they would complete enough work on their grounds by the end of June to be able to stage gigantic, and rival, Fourth of July picnics.

Much excitement and speculation surrounded the marriage of President Grover Cleveland to his 21-year-old ward, Frances Folsom, on June 2, 1886.

It was an exciting summer, not only in Dallas, but all across the United States and Europe. National attention focused on Washington as President Grover Cleveland married his 21-year-old ward. Mad King Ludwig of Bavaria drowned mysteriously. Steve Brodie jumped off the Brooklyn Bridge and not only lived to tell about it, but made the tale profitable.

Rumors from Russia speculated on the abdication of Prince Alexander. Thousands died when Charleston, South Carolina, was devastated by a severe earthquake. And in New York City, preparations were being made to dedicate the Statue of Liberty.

Construction at both fair sites continued into fall. The absurdity of separate and simultaneous expositions was now accepted with a certain amount of good humor and treated as a unique illustration of the city's prosperity and independent spirit. New attractions and exhibits were announced daily including several previously displayed at the 1884 World Industrial and Cotton Centennial Exposition in New Orleans.

Expressions of statewide interest boosted hopes that as many as 200,000 visitors might come to Dallas for the two events, and local businesses stocked up for the expected increase in trade. Committees met in homes and hotels throughout the city. Among the many meetings, one was called to discuss ways Christian ladies could make money at the fairs to benefit a Home for Fallen Women which was located in Ennis.

Privileges, the rights for commercial booths on the grounds, were auctioned to the highest bidders during the first week in October. Individuals, churches and social organizations paid for the opportunity to sell such items as ice cream, watermelon, hot candy and cigars, or to operate the photo and shooting galleries, barber shop and various tented shows. The privilege to manage the bar under the Dallas State Fair grandstand sold for $200, while the raw oyster booth was a relative bargain at $112.50.

At last, under a cloudy sky that threatened rain, the Texas State Fair opened as planned on October 25, one day ahead of its competition. Visitors, reportedly about 3,000 over the course of the day, traveled by rail or private vehicle to the grounds north of Dallas, where they discovered construction still in progress and exhibits not yet in place. The upper floor of the exposition's main hall remained closed to the public, and those who had expected to see the much ballyhooed taxidermic exhibit, a collection of nearly 100 stuffed birds entitled "The Burial of Cock Robin," were disappointed.

Prior to the governor's speech at 4 p.m., fair officials announced they would be unable to present the entire first day's program and no competitive awards would be made. A scheduled balloon ascension by Professor Brayton, advertised to feature heart-stopping stunts on trapeze and rings, was cancelled because of high winds. As darkness approached, the

Fairs are as American as flags, moms and apple pie — which not surprisingly can be found in abundance at most fairs. There are more than 3,000 fairs and expositions in the United States each year, and total attendance for these events tops the 150 million mark. That's more than twice the combined annual figures for major league baseball, pro football and New York theaters.

America's first fair was held in 1804 in the nation's capital with prizes awarded for superior produce and the best farm animals. A few years later, Parke Custis, a stepson of George Washington, organized an annual sheep-shearing contest in Arlington, Virginia.

The tinkling of a sheep's bell signaled history in the making in Pittsfield, Massachusetts. With show-manship flair, a country gentleman named Elkanah Watson laid the foundation for the American agricultural fair industry. Because his neighbors had never seen his pair of Merino sheep, Watson herded the shaggy twosome onto the village green to graze, and curious towns-people stopped to ask questions and admire the fine wool. Watson expanded this concept when he organized the first Berkshire Cattle Show in 1810. Next he added cooking contests. But competition was considered "unladylike," and women were reluctant to accept their awards in public.

Syracuse hosted America's first state fair in 1841, and county fairs sprang up in many parts of the country. An aura of anticipation surrounded these annual events. The McCormick Reaper, John Deere's steel plow, threshing machines, corn planters and cultivators were exhibited. Both Cyrus Curtis of publishing fame and the legendary Henry Ford said they were inspired toward their remarkable careers by the new machines they saw at fairs.

The London Society of Arts presented an "Exposition of the Industry of All Nations" in 1851. This first world's fair was housed in the magnificent Crystal Palace, a vast temple of glass and iron that covered 23 acres.

A smaller structure of similar design was built for the first major exposition in the United States. The New York Crystal Palace Exhibition of 1853 featured the new passenger elevator and sewing machine demonstrations. In 1876, the Philadelphia Centennial Exposition marked the nation's 100th birthday and introduced Americans to such wonders as the telephone, typewriter and bicycle.

chariot races, conducted as promised in full Roman costume, finally got underway, but by this time most of the fairgoers had gone.

It was an inauspicious start for the six-day run. The directors were further embarrassed to learn that a traveling circus, set up on a lot downtown, had drawn considerably more persons to its two performances that day.

Meanwhile, at the Dallas State Fairgrounds, throughout the day and into the night, workmen hammered and painted, exhibitors finished setting up, and last minute landscaping gave an illusion of permanence to the site.

Hundreds were already camped outside the gates. And while fair secretary Sydney Smith and Tom Marsalis, the director in charge of grounds preparation, worked late as usual, Billy Gaston made a final check of the livestock areas and went home for a family celebration. It was Gaston's 46th birthday.

Sydney Smith served as the fair's secretary from 1886-1888 and again from 1896 until his death in 1912.

1886

From the opening minutes until the last streetcar left at midnight, there was no doubt about the public's enthusiastic response to the Dallas State Fair. First day attendance was 14,000 — men in bowlers and spats, women in bustled gowns of silk tafetta, children on their best behavior. The crowd was even larger the second day and reached 20,000 on the third.

While the daily racing program was unquestionably the top attraction, other features, particularly those seldom or never seen previously in North Texas, drew crowds and plaudits.

Bookmaking and pool selling at the State Fair racetrack.

Traction engine displays were featured at the first State Fair.

Visitors to the Main Hall were introduced to new products and viewed top-of-the-line everyday goods in imaginative settings. The Dallas Soap Works exhibit was a miniature cottage constructed entirely out of soap from ladies' fine toiletries to the washtub variety. A New Orleans manufacturer presented a demonstration of artificial limbs in motion. Linz Brothers of Sherman displayed a $25,000 diamond necklace, and Liggett & Meyers Tobacco Company of St. Louis recreated an oldtime log cabin with tobacco growing in the front yard, a live black "auntie" busy with chores inside and "Uncle Ned" playing a fiddle on the porch.

The featured item in the ladies' department, which was located in the gallery of the Main Hall, was a unique piece of handwork. Miss Mamie Hereford of Dallas had etched her epic poem, "The Land of Dixie," in bright gold letters on a nine-foot banner of sapphire blue satin. The poem had six verses and contained 1,100 individual letters.

Along with tools and farm equipment, the Machinery Hall showcased the textile and garment industry by displaying all the machines necessary in the process to convert cotton into thread. A section of a clothing factory was set up, and 20 operators labored during regular fair hours. Mechanical luxuries, such as an ornamental lawnhouse with revolving doors, were also in evidence.

Outdoor entertainment included a concert presentation of "Il Travatore" by the Mexican National Band with anvil accompaniment furnished by 25 members of the Busch Zoave and a spectacular fireworks show which featured elaborate set pieces. One of these depicted a man crossing the Trinity River and being engulfed by a rising flood of silver fire. Another evening's highlight was a grand war dance staged by 100 Comanche Indians under a shower of electric light. The Indians, something of a novelty to area residents, made several appearances during the fair and tolerated head-to-toe inspections by Texans whose curiosity had been stirred by the exploits and recent capture of Geronimo.

A registered Hereford show, described as the most important ever held in the state, drew 18 entries, although a couple of purebred Brahman on display received the most attention. Livestock shows and auction sales provided local producers with an opportunity to upgrade their own herds after examining the various breeds of beef and dairy cattle, hogs, sheep and goats.

The Dallas State Fair closed on November 7. Estimates of total attendance ranged from 100,000 to 250,000, and Alex Sanger, speaking in behalf of the weary directors, commented: ". . . we are well satisfied, even if we have not made much money." Fair officials proudly claimed that the exposition was responsible for opening up miles of Dallas streets and putting at least $750,000 into circulation.

Opening Day at the Dallas State Fair drew a crowd of 14,000.

On a smaller scale, the Texas State Fair, which had rebounded after its shaky opening, was also adjudged a triumph. Together, the two fairs had produced a revenue bonanza for barbers, livery men, saloon keepers, hack and express drivers, newsboys and grocers. Restaurants located on the main lines of streetway travel to the fairs enjoyed brisk business. Large hotels and boarding houses made money on meal tickets, though less than anticipated on accommodations, while smaller and second class establishments prospered in all areas. Dry goods merchants reported little effect on sales, and Dallas lawyers, bankers and brokers considered the period a pleasant, but not profitable, holiday. To the city's credit, there were no reports of price gouging.

In this aftermath of success, voices were heard again urging that the rival organizations overcome their differences and unite to produce one great state fair the following year. Popular acclaim and community benefits aside, receipts from the Dallas State Fair fell more than $100,000 short of meeting expenses. Both sides were ready for compromise.

The agreement came on February 10, 1887. Hardware merchant James Moroney was elected president of the consolidated enterprise which would be presented on the East Dallas site. Styled as the Texas State Fair and Dallas Exposition, the new corporation began with limited cash, sizable debt and high hopes, based in part on its distinction as the only fair in town.

Uphill All the Way

1887-1904

1887

James Moroney, State Fair President, 1887 and 1906.

Good times continued to get better in the late 1880s. It was an optimistic, flamboyant era. Money-making was on the minds of most Americans. Transcontinental railroads spurred the development of powerful national industries. Retail selling, once the province of peddlers and general stores, was becoming big business.

As westward expansion drew to a close, cities replaced frontier towns, final battles were fought in the Indian wars, and long cattle drives gave way to the efficiency and economy of rail transportation.

Texas experienced phenomenal growth and change. Nowhere was this more evident than in Dallas, a young city which could trace the reason for its prosperity directly to the railroads, a city which would soon surpass San Antonio, Galveston and Houston as the state's largest.

In 1887, buildings were going up in Dallas at twice the rate of the previous year: 81 business houses, 312 residences, three churches, one school and 300 additional homes in the suburbs. The Farmers' Alliance, an agrarian union that was gaining political strength in Texas, moved its state headquarters to Dallas and erected a handsome marketing and purchasing warehouse. Other major construction projects included a new city hall and a $250,000 cotton mill, the first in this area and vitally important to manufacturing interests in Dallas.

Those who weren't building were buying and selling — 5,784 pieces of property changed hands over the 12-month period. The development of Oak Cliff easily qualified as the year's most ambitious real estate venture. This 2,000-acre suburban addition on the far side of the Trinity River was the brainchild of Tom Marsalis and his partner in the wholesale grocery business, John S. Armstrong.

In the wheeling-and-dealing commercial atmosphere of 1887, the directors of the newly-reorganized Texas State Fair and Dallas Exposition chose not to concern themselves unduly with the $100,000 holdover debt, but moved quickly to acquire 37 acres adjoining the western boundary of the original fairgrounds. They purchased two tracts with cash, and Billy Gaston, owner of the third, accepted a five-year note as payment.

Construction began on a Geological Hall near the vehicle entry to the grounds and a new Machinery Hall which would provide 90,000 square feet of exhibit space. The Dallas Jockey Club elected to build a clubhouse near the grandstand for its 70 members, and several manufacturers decided to erect their own display buildings in an area designated as Machinery Park. Stalls were added to boost the fair's livestock capacity to 1,500 horses and cattle and 500 sheep and hogs.

Governor Sul Ross announced he would come from Austin for the opening, and with prospects of a significant increase in out-of-town visitors, a committee was formed to assist with accommodations. By opening day, lodging had been located for 8,143 persons at daily rates ranging from 25 cents to $1.50. An additional 5,000 were expected to stay with relatives each day.

Those traveling to Dallas expected an even bigger and better show than before, and they were not disappointed. Eye-openers ranged from a model of the Washington Monument constructed entirely out of human teeth, to an enormous roller coaster patterned after the first gravity railway built at Coney

An award of special merit presented in 1887.

Directors of the Texas State Fair and Dallas Exposition voted to expand their property to 117 acres with the purchase of three parcels of land along the western boundary in 1887.

Island three years earlier by LaMarcus Thompson. There were panoramas — living enactments of the Battle of Gettysburg and Custer's Last Stand; 15-pound yams and beets 30 inches in diameter; new inventions such as a machine for killing fish electrically and another that washed and dried dishes.

The livestock department offered 2,000 conventional entries plus a drove of Texas-raised camels and a special pet section featuring Carlo the Rescuer. Carlo was a locally-famous dog, a large Newfoundland that had saved a woman from drowning. An engraved medal attested to his courage, and Carlo was a popular attraction until the day he got loose and killed two white rats and a premium mockingbird before being captured while in the act of devouring a prize-winning cockatoo.

In addition to the primitive coaster, fun seekers found flying jennies, which were forerunners of the merry-go-round, Punch and Judy shows, skill games such as ring-over-a-cane toss and sideshows which headlined, among others, Jack the Giant and a mermaid described as having a "lovable looking upper torso and the lower body of a speckled trout." The man in charge of the mermaid started a new show every three minutes and raked in dimes all day long.

Hungry fairgoers discovered restaurant row. The sounds and aromas of food preparation hung over the crowd as people

The vehicle entrance to the fairgrounds.

pushed and edged their way along the street from one booth to the next. Lost children, pitchmen and pickpockets mingled together, and fist fights settled minor differences. Police routinely rousted poker games in establishments along the row.

A spectacular balloon act was billed as one of the fair's main attractions. Professor Gilbert handled the hot air mechanics, while his beautiful daughter electrified audiences by parachuting to the earth from a height of one mile. Prior to the first performance, a reporter for the *Dallas Morning News* approached Gilbert and asked to accompany the girl. Already beset with technical difficulties and poor atmospheric conditions, the professor turned him down. Undaunted, the reporter waited until seconds before take-off, then ran and leaped in beside the surprised stuntwoman. The two began a wobbly ascent — Miss Gilbert smiling and waving to the crowd, the reporter crouching in the bottom of the basket. At 500′, the extra weight cracked the wooden hoops around the balloon's neck. With the framework splitting, canvas flapping and Miss Gilbert still waving, the balloon sailed across the racetrack and just cleared the domes of the Main Exposition Hall before dropping abruptly onto the Texas & Pacific tracks and ripping apart. No one was hurt; Professor Gilbert found another balloon for the act; and the first-person story ran on page four.

Another highlight of the fair was the Wild West Show, a rodeo-type event, staged inside the circle of the racetrack. Wild cattle were penned in the center of this area, then released to be chased and roped by cowboy contestants. Concerns about public safety were handled in the direct and logical fashion of the times. Four men on horseback were stationed around the track with rifles and given instructions to shoot any longhorn that jumped the fence.

The Grand Baby Show was scheduled on the final day of the exposition. A $50 prize was offered in a competition for toddlers and infants, the winner to be determined by votes cast on that day's admission tickets. Hoping for a good response, fair officials set aside a hall outfitted with 15 baby swings. By 11 a.m., the room was packed with 107 crying entrants, an equal number of proud mothers and assorted friends and relatives lobbying for their favorites. Voting took four hours, first on printed tickets, and when these were gone, on slips of paper. Recognizing an impossible situation, the beleaguered officials finally locked the ballot box and announced there would be no count and no winner until the following Monday.

Cumulative attendance figures were not recorded during the early years of the Texas State Fair and Dallas Exposition, but from the total receipts it appears that the second fair drew about the same number of people over 17 days that the first

In common with many former Confederate officers, Sydney Smith used his army rank as a title for the remainder of his life. The fair's first secretary had served as captain and aide-de-camp to General Stonewall Jackson. A graduate of the University of North Carolina, Smith tried cotton farming in Mississippi after the war before moving his family to Dallas in 1878.

Rail connections made Dallas an ideal location for northern and eastern manufacturers to establish distribution centers, and Sydney Smith found his niche in the business world as a local manager for various out-of-state companies. He was thereafter a great champion of the traveling sales representatives, known as "drummers," and pushed for the establishment of Drummers' Day at the State Fair.

fair did in 12. For whatever reasons, when all the money was counted and all the bills paid, there was another deficit amount of nearly $17,000 to be added onto the directors' personal notes. Aside from the fact that it didn't make money, the fair exceeded expectations, and the stockholders voted to extend the next year's exposition to 21 days.

In September of 1888, international news centered around the Eiffel Tower, nearing completion in Paris, and a horrifying series of murders in London committed by a still-at-large criminal nicknamed "Jack the Ripper."

In Dallas, people were looking forward to another fair and a better racing program. Some of the country's finest flyers and trotters had been shipped to Texas following the prestigious St. Louis Fair. There was no need to stimulate enthusiasm among potential exhibitors; in fact the demand for space so outdistanced the supply that the "haves" were implored to double up and share with the "have-nots."

John S. Armstrong, who had extricated himself from his real estate partnership with Marsalis, was now president of the association. Sydney Smith continued as secretary, in effect the general manager of the operation, and his wife, Margaret, again assumed duties as superintendent of the ladies' department.

The 1888 exposition started off with resounding explosions from each of three cannons positioned around the outskirts of the city. This 21-gun salute, fired at 6 a.m. to make sure everyone would be up in time to see the parade, was in honor of the 21 Texas counties bringing exhibits to the fair.

The livestock department had been relocated from the back of the grounds to an area not far from the Main Exposition Building, a move which proved unpopular with exhibitors who complained constantly about the flies and odors emanating from the stalls.

Several smaller structures had been built during the year, and one of these was designated as the "colored people's department" and was reserved for displays organized by Negro churches and associations. A steady stream of visitors, blacks and whites, flowed through this department during the fair.

Opening day festivities were marred by a bloody accident on the giant coaster. The ride was constructed with parallel tracks running an undulating course between two towers. Each car, carrying up to 12 passengers, swooped down from the station, gained momentum which carried it up and over several inclines until it came to a stop and locked on the grade leading to the opposite tower. The passengers would get out and walk up the edge of the track while attendants retrieved the car, pushed it up to the platform and then switched it to the other track. The passengers climbed back in, and the whole process began again. On this night, however, an inexperienced attendant forgot to switch the car, instead sending it careening back down the same track head-on into the path of a second car. Five people were seriously injured with broken bones, deep lacerations or internal bleeding, and one man died the following week. The coaster operator wasted no time with explanations; he simply disappeared.

The fair's featured musical attraction was Liberati's Band, led by the Italian-born cornetist who was reputedly the handsomest musician in America. The band presented two or three concerts daily.

John S. Armstrong, State Fair President, 1888 and 1890.

Inside the Main Hall, fairgoers enjoyed "The Wonder World," a 32,000 square foot exhibit which used 500 moving figures to illustrate scientific discoveries. The center of the display was devoted to Darwin's theories on the origin of man, a subject apparently not considered too controversial for public viewing.

Elsewhere in the building, businessmen inspected Edison's new mimeograph machine and the latest model typewriters, while ladies took turns giving impromptu recitals on the pianos and organs in Watkins Music Company's display. The Singer Sewing Machine exhibit showcased an etching of "Our Uncrowned Queen," the now 23-year-old Mrs. Grover Cleveland.

Those looking for sights of a less serious nature wandered outside among the sideshow tents for a glimpse at the 850-pound fat man or the fellow who advertised his eight-foot beard as the world's longest.

The most-anticipated event, a speech by the noted southern orator Henry W. Grady of Atlanta, produced the largest one-day attendance of this exposition or its predecessors. People began arriving at 8 a.m. and quickly filled the 7,000 available seats in the grandstand with another 3,000 overflowing into adjacent areas. With the spectators becoming restless, Liberati's Band attempted to entertain the gathering with operatic airs, but the crowd demanded and finally got a rousing rendition of "Dixie." Grady's oratory addressed the Negro question in the south, and he passionately advocated that ". . . people once an element of wealth in their subjection become an element of wealth in their freedom."

The baby contest on the closing day was not without controversy. One infant, entered as an orphan in need of money, received a large sympathy vote and won first prize.

Bandleader A. Liberati.

After complaints were raised that this child had two healthy parents, an investigation revealed that the winner was an adopted orphan.

The directors and stockholders were well satisfied with the 1888 celebration. The fair still hadn't made money, but it hadn't lost as much either. Another $8,000 was added to the accumulated debt.

1889

There was big news north of the Red River in the spring of 1889 when the federal government opened the Indian Territories for settlement. The subsequent land rush created tent cities of 10,000-15,000 inhabitants, literally overnight, in Guthrie and Oklahoma City.

Dallas celebrated the completion of a new city hall and watched a rivalry brewing between its major hotels — the seven-story McLeod, later known as the Imperial, and the expanded, by two stories and 200 rooms, Grand-Windsor.

Not everyone was sharing equally in this period of growth and prosperity, however. The same factors that fed the rapid development of cities like Dallas worked corollary hardships on surrounding rural areas. The Farmers' Alliance suffered a severe setback when its huge warehouse was forced into bankruptcy and had to be sold after only 20 months of operation.

Henry Exall, State Fair President, 1889.

The Texas State Fair and Dallas Exposition was also caught in a financial squeeze. With local banks unwilling to carry the organization's increasing indebtedness, now totaling $124,000, it was necessary to secure bonds using the property as collateral. When the Holland Trust Company of New York agreed to issue bonds for no more than $100,000, directors Sanger, Gaston, Armstrong, Marsalis, Blankenship and Simpson reached into their own pockets and paid off the remainder.

Henry Exall, known for the fine racehorses bred on his farm north of town and for his active interest in improving agricultural production, was named the fair's new president. C.A. Cour was appointed as secretary.

The exposition was scheduled for 13 days, as management continued to experiment with the length of the run. Secretary Cour booked the renowned Flying Artillery from San Antonio and Cappa's 50-piece Seventh Regiment Band of New York City. The featured attraction was an Irish setter from Canada named "Doc."

Upon his arrival in town, Doc received the red-carpet welcome due a racing champion. Counterparts of modern-day groupies appeared at the St. George Hotel where the celebrated canine and his trainers were staying. Doc's reputation had been built at the expense of ponies and cyclists. Hitched to a sulky and competing as a trotter over a half-mile track, the dog was virtually unbeatable, as he proved against a series of challengers throughout the fair.

The popularity of the Flying Artillery reflected an ongoing interest in military weapons and manuevers even though there had been no deployment of U.S. troops for nearly 25 years. The artillery program showcased the capabilities of horse-drawn heavy cannons. During one such drill, 200 people climbed out on the roof of the Main Exposition Hall to get a better view. The weight caused a 25′ section of the roof to collapse. None of the spectators fell through, nor was anyone below hit by debris, but roof repair became a priority on the directors' list for off-season improvements.

As always, fairgoers were treated to some unusual sights. A Tyler man displayed his multi-purpose invention designed to simultaneously churn butter, rock the baby's cradle and keep flies off the table. An exhibit sponsored by the Sanger Brothers store told the story of Rip Van Winkle using wax figures in a setting with real water trickling over rock-covered hills.

Along with the usual contests to determine the best pair of Cayuga ducks or the blue-ribbon jar of pickled cucumbers,

Opening Day attendance in 1889 appeared to double that of the previous year.

urban curiosities were whetted by a competition among three kinds of road machines to establish superiority in the practical art of grading streets.

The delightful Indian summer temperatures prompted a rumor, quickly denied, that the exposition would run an extra three days. It was perhaps the only time that anyone would ever suggest the fair be extended because of good weather.

Though well-received and bathed in sunshine, the 1889 fair still failed to achieve the elusive balance between income and expenditures. In December, a second mortgage of $50,000 was placed on the grounds with 22 companies and individuals, including most of the indefatigable directors, subscribing for this amount.

1890

The 1890s dawned with industrialization and urbanization as accepted facts. The passing of the frontier was mourned by many, among them Theodore Roosevelt, who feared that the cherished values of courage, independence and self-reliance would be threatened by the new order.

Indeed, attendant problems were running on the heels of progress — power abuses by the railroads and utility companies, business monopolies, political corruption, rising costs of manufactured goods and declining prices of agricultural commodities.

For the first and only time on record, the U.S. Census ranked Dallas as the largest city in Texas, an honor assured by the annexation of East Dallas and its 5,000 inhabitants. According to the 1890 count, Dallas County was also the most populous in the state. Oak Cliff was separately incorporated

In 1890, a Music Hall with seating for 3,000 patrons was constructed as an addition onto the northeast end of the Exposition Building.

with 2,470 residents, and its leaders hoped that the new city might someday relate to Dallas as Cambridge to Boston or Brooklyn to New York City.

John S. Armstrong accepted a second term as fair president, and a large Music Hall was built adjoining the northeast end of the Exposition Building. The new addition had a seating capacity of 4,000, and was used continuously throughout the fair for band concerts, performances by a troupe of educated horses and dogs, pipe organ recitals, and special shows by Stewart the male soprano, a female impersonator who delighted audiences by wearing a different costume every day. Mr. Innes, leader of the featured band and self-acclaimed as the world's greatest trombone player, objected to following the horse and dog act onto a sawdust-covered stage, so his concerts were shifted to outdoor venues whenever possible. Midway

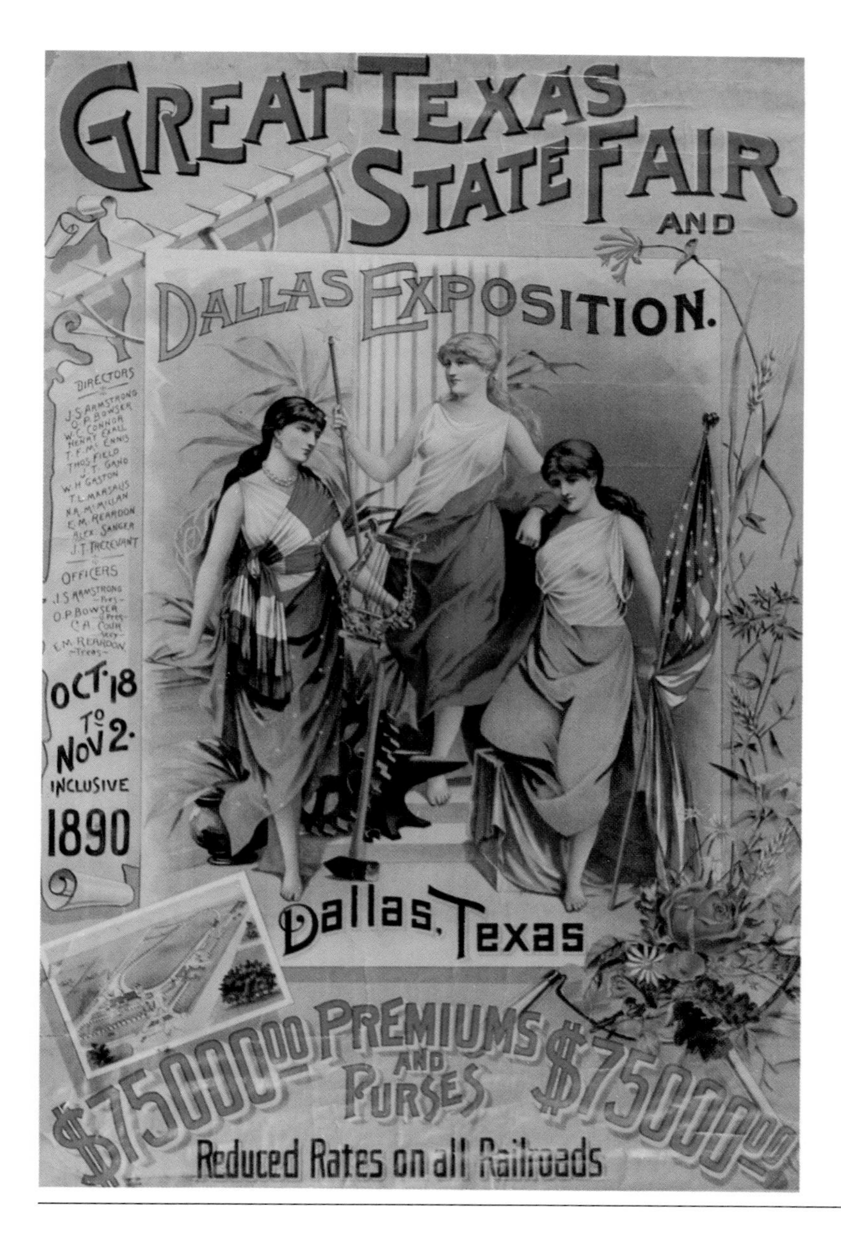

Missouri's governor David R. Francis delivered the Opening Day address.

through the fair, the new facility was closed down, cleaned up and reopened the same evening with a beautiful tinted canopy erected inside to create an elegant setting for the seventh annual Idlewild Ball, the premier event of the Dallas social season.

Other structural improvements included an electric railway which visitors could ride from one activity site to another and a fresh coat of paint on all the buildings. A circular swinging ride called the Razzle Dazzle was offered for the first time, and a lost children's station was established in the Main Hall. Stray youngsters were delivered to an attendant who rang a loud, clanging bell until the right parent realized someone was missing.

Among the guest speakers at the 1890 exposition were the governors of Missouri and Tennessee, the latter providing the highlight of Tennessee Day, which attracted what appeared to be a record crowd, estimated between 25,000-35,000.

Wide-eyed visitors wandered through a Japanese Village staffed by eight nationals who explained the mysteries of the Orient, and they stopped at the skull collection of Dr. Windsor, a phrenologist who demonstrated how the head of a murderer could be recognized from the peculiar arrangement of bumps and ridges. The swine show was larger than ever, a testimony to the importance of huge refrigerators and packing houses recently erected in Texas.

Admission to this myriad of sights and experiences was still only 50 cents, but anyone arrested for scaling the fences was hustled off to trial and if convicted, given the alternative of an $18.50 fine or 36 hours working on a chain gang.

A new promotion was tried at the end of the run — Red Headed Day. Carrot-topped kids and red headed people riding white horses or mules were admitted free. Gold medals were awarded to the handsomest and ugliest red heads. Though colorful, the event did not have the broad appeal of the baby contest and was not repeated.

By every indicator save one, the 1890 fair was a tremendous success; but as usual costs were greater than income, in this instance by slightly more than $26,000.

1891

Any hope of reversing the trend was dispelled the following spring when the entire complex of racing stables burned to the ground. The stables were not insured, and the loss was set at $13,000. To rebuild, new president W.C. "Bud" Connor and the directors signed a note to Elliott Lumber Company secured by the buildings at the front of the grounds, the association's only assets not already mortgaged.

As the fair began the task of replacing 400 stables, construction was nearing completion on a $300,000 Dallas County Courthouse built out of red sandstone and the city's newest house of worship, First Baptist Church. These magnificent Romanesque structures survive today as examples of Dallas' early downtown architecture.

Speculation about the twentieth century was a popular pastime. Editors of the *Daily Times Herald* looked into the future and predicted that in 100 years electricity would power societies. They also foresaw great airships which would transport passengers and freight around the world in ten days or less. By 1991, they opined, the population of Texas would be larger than that of India.

W.C. Connor, State Fair President, 1891.

The Dallas business community rallied in support of the 1891 fair. Many organizations purchased admission tickets to give to preferred customers, and 122 businesses closed to encourage opening day attendance.

Governor Hogg came from Austin for the occasion and noted that the State Fair's opening was always blessed with beautiful weather. This remark promoted a fire insurance agent to counsel, ". . . some houses never burn, but they get nearer to the day they will burn."

Liberati's Band was booked for a return engagement and provided a model for the Sangers' display which used 150 dolls to recreate a concert scene. The department store offered to give the dolls away after the fair. A box was set up near the exhibit and votes cast for 10 cents each to decide the winners. A day-by-day tabulation of the ballots ran in the newspapers, making possible an eleventh-hour avalanche of votes which assured victory for children from several prominent families.

Every afternoon, fairgoers watched intently as Professor Snyder walked a rope strung between the domes on the Exposition Hall. And they were equally fascinated with Millie Christine, billed as the "two-headed woman". In addition to her two heads, Millie Christine had four arms, four legs, but only one shared torso. Spectators marveled that she could speak French or German with either mouth and sing alto and soprano at the same time.

An exhibit featuring relics of South Sea cannibals appealed to the primitive instincts of many visitors as they lined up to look at instruments identified as head splitters and skull crushers along with assorted blood cups and cooking bowls.

Stock in the World's Columbian Exposition, scheduled to open in Chicago in 1893, was sold from a booth on the grounds. There was much interest in this upcoming World's

Siamese twins Millie and Christine, billed as the two-headed woman in 1891.

A SONG WITHOUT MUSIC

I am going to the Fair,
In the Fall;
I will try to enter there
On my gall;
If I haven't got the pence
I will get in through the fence
And they cannot drive me thence,
Not at all.

I would rather live on hay,
So I swear,
Then have to stay away
From the Fair;
If the wealth is not around
I will ride in on a hound,
I will burrow underground
Getting there.

— author unknown

Fair, particularly with regard to Texas representation. Many exhibitors felt federal appropriations for the event should be used in part to provide an area for exhibits from Texas. The World's Fair Association insisted that Texas would have to pay $100,000 to erect its own building. Without settling this question, tentative plans were made to spend the next year collecting materials which first would be displayed at the fair in Dallas and then shipped to Chicago.

Final attendance and receipts were comparable to the 1890 exposition, and the year-end financial picture was now grim. In addition to the bonds, the fair owed $66,226, which represented money due to premium winners, unpaid notes and advertising debts. Jules E. Schneider took over as president of an association that could no longer obtain credit. Because of the debts, related suits and the fair's failure to meet interest obligations, the Holland Trust Company demanded payment on its bonds.

Total organizational collapse and loss of the property was avoided only by structuring a new corporation. Holders of the second mortgage bonds agreed to cancel this debt. More personal guarantees were offered by individual directors, and through a series of transfers and sales, most of the original bonds ended up in the possession of the Manchester Trust Company of England.

1892

Jules Schneider, State Fair President, 1892

Schneider then began to prepare for the 1892 event using his own personal credit whenever borrowing became necessary. He refused any salary and recommended that future presidents consider their service a contribution to one of the state's most important resources.

No one disputed the exposition's value to Dallas in terms of the thousands of dollars spent locally by out-of-town fairgoers. Many of these visitors were so impressed by Dallas that they eventually moved to the area. Others relocated businesses to Texas or invested money in Texas-based operations. Livestock bloodlines had been improved through fairtime sales, and nationally-known performers brought an element of cultural sophistication to the city. Those businessmen who gave so generously to keep the exposition alive were keenly aware of the direct and indirect benefits that accrued from having the State Fair in Dallas. The altruistic motives behind their efforts did not obscure practical considerations.

News headlines that fall announced the upcoming presidential election and the soon-to-be-dedicated World's Columbian Exposition in Chicago. The death of popular poet

John Greenleaf Whittier was noted, and Americans breathed a communal sigh of relief when the Dalton Gang was gunned down during an attempted bank robbery in Coffeeville, Kansas.

Two days before the opening of the 1892 Texas State Fair and Dallas Exposition, it began to rain. Heavy showers fell on North Texas for 48 hours. Although the sun made a brief opening day appearance, the cycle clubs bowed out of the parade which was routed over streets reeking with slush and mud, and the races were called off for fear of endangering the horses.

Rain continued, forcing the cancellation of artillery drills, balloon ascents and the entire first week's racing program. An announcement was made that the football game between Dallas and Fort Worth on Columbus Day would be played regardless of weather conditions, and the rain came down in torrents.

Indoor activities drew respectable crowds. Madame Marie Decca, known as the American Jenny Lind, was the featured entertainer along with Liberati's Band and the Royal Theater Company Marionettes. Visitors filled the Exposition Hall galleries admiring the art and handwork. They spent money at exhibit booths such as the one displaying rattlesnake skin neckties, alligator toe purses and wild animal pelts to be used for rugs and robes.

The directors vehemently denied a report in the *Fort Worth Gazette* that the fair would be extended. They laid plank walks

An impressive display of home-canned fruits.

from the entrance to the Main Hall and other points of interest and ran ads stating that the exposition positively would not run past October 30.

Cold, gloom and constant drizzle washed out the Drummers' Day parade, but Confederate Reunion Day brought out 6,000 veterans who scoffed at the weather saying they certainly had endured worse during the war.

With five days remaining, the sun broke through, and the first races were run in admittedly heavy footing. Attendance picked up dramatically, and management, sensing an opportunity to recoup some of the lost racing revenue, reversed itself and announced that the fair would be extended to November 4. They also decided to select the best of the county exhibits and send them to Chicago for a collective display at the World's Fair. A committee was appointed to raise $25,000 to transport the materials and build an annex to house them near the existing Texas Pavilion.

Then the clouds reappeared. A pouring, pelting rain fell for two days turning the track into a hopeless morass of mud. Finally, on the morning of November 1, Secretary Cour met exhibitors, privilege men and visitors at the gate and on behalf of President Schneider, who was home in bed with a terrible cold, explained simply, "The fair has closed."

The fair and its problems were forgotten for the moment in the excitement surrounding national and statewide elections. Voters returned Grover Cleveland to the White House and endorsed James S. Hogg's bid for a second term as governor of Texas.

At the annual association meeting in December, stockholders learned that the fair had managed to pay its running expenses, settle some past obligations, and while unable to reduce the bonded debt, would finish the year with a cash balance of $10.93. John N. Simpson was elected the association's new president.

1893

Simpson's background as a Confederate colonel, cattleman and successful banker made him an ideal leader in the financially troubled year of 1893. Even before the country's first rash of railroad and business bankruptcies and the resultant market collapse, local doomsayers expected hard times for the 1893 Texas State Fair in its head-to-head competition with the spectacle taking place in Chicago.

The World's Columbian Exposition, which commemorated the 400th anniversary of Columbus' discovery of America, provided the nation with one shining achievement in a year which saw 500 banks and 16,000 businesses fail, unemployment touch 20% of the labor force and panic deteriorate into a lingering depression. The six-month event attracted 21 million visitors to a 666-acre exposition park laced with canals and lagoons and illuminated with more electricity than was being used by the entire city of Chicago. A plaster and fiber finish on the main buildings gave the appearance of white marble in the sunlight. Inside the gates of this classically-designed "White City," sightseers were introduced to such diverse wonders as the linotype machine, Pullman car, Ferris wheel and a scandalous new kootch dance performed by the alluring Little Egypt. Chicago was only 30 hours from Dallas by train, and the round-trip fare was a reasonable $28.95.

Five Dallas banks folded before the year was over. Thomas Marsalis lost all his investments and properties in Dallas County, and Tom Field, unable to complete construction on the opulent $500,000 Oriental Hotel, was forced to sell his interests to a group of St. Louis investors.

Under the circumstances, there was minimal demand for exhibit space at the Texas State Fair. The directors voiced their belief that the fair would completely obliterate the effects of the panic in Dallas by drawing 140,000 outside visitors. Then they used their personal leverage in the community, twisted a few arms and filled the halls with exhibitors.

There were indications, however, that even the directors were apprehensive. They reduced the size and number of newspaper ads, cut expenses in the entertainment area and for the first time actively sought Fort Worth participation.

A model of the U.S. Cruiser Charleston, with a midget admiral and three beautiful girls on board, and Sampson the Strongman were two of several exhibits acquired from the World's Fair. Another highlight was the comprehensive

John N. Simpson, State Fair President, 1893 and 1919.

Entries line up for the start of the 1893 parade.

In the spring of 1893, Dallas celebrated an event that was expected to prove as significant to the city's future as the arrival of the railroads 20 years earlier. The longstanding challenge of creating a shipping channel to the Gulf via the murky, capricious Trinity River had been accepted by the Trinity Navigation Company. On May 24, after an obstacle-filled 67-day trip up the river from Galveston, the "H.A. Harvey, Jr." was welcomed into Dallas by a huge crowd. The steamer operated for several years, primarily making excursion trips to a point 14 miles downriver, but the original project lost momentum as the economy sagged, and the "Harvey" was finally sold in 1898.

Texas exhibit prepared for, but never sent to Chicago. In all, the interest in the Columbian Exposition and the slumping economy combined to benefit the State Fair, since many Texans were unable to travel north and chose Dallas as an affordable alternative.

One of the fair's new features was a woman's congress, organized by Mrs. Sydney Smith to provide opportunities for ladies engaged in professional and industrial pursuits to exchange opinions and experiences. Scholarly papers were read on such pro-suffrage topics as "Equality, Not Supremacy," while disenchanted male bystanders snickered that, "one mouse would disrupt the whole proceedings."

The weather was as consistently clear and sunny as it had been wet and dismal the previous year, contributing to an exceptional race meet that was enlivened by a minor scandal. One of the early winners, "Little Fred," turned out to be an Illinois champion entered under an assumed name against much slower competition. Gossips also reveled in the news that Madame Decca, star performer from two years earlier, had developed a drinking problem and was being divorced by her husband.

During the second week, the railroad lines advertised a special one cent per mile rate from anywhere in Texas. This offer along with the good weather produced a tremendous response. Pessimistic attendance predictions gave way to complaints about packed street cars and long lines at the entry gates. There was speculation that the exposition might make money.

President Simpson confirmed the rumor. For the first time in eight years, the fair had operated at a profit. Simpson paid debts from the preceding year and the current bond interest before turning over a small amount of cash to his successor, Alex Sanger.

1894

Although the economic malaise prompted by the nationwide depression had settled fully over Dallas in 1894, Alex Sanger and the other fair directors, buoyed by the organization's first signs of financial stability, moved to secure new attractions popularized at the Columbian Exposition. For all its architectural and scientific achievements, the recent World's Fair owed its mass appeal to the exotic entertainment presented on the Midway Plaisance.

The Texas State Fair announced its own enlarged midway for 1894 with features direct from or equal to those shown in

Alexander Sanger was the youngest of four German-born brothers who pioneered the mercantile business in Texas by opening a chain of stores along the expanding Houston & Texas Central Railroad line. Sanger came to Dallas in 1872 to set up the family dry goods operation in a two-story brick building on Main Street. He was joined by his brother, Philip, and together they built a retail establishment known for its quality merchandise and innovative services. Sangers was not only an immediate commercial success, it quickly achieved status as an arbiter of taste in the young city by establishing a standard for other fine stores and encouraging the parallel development of the fashion industry in Dallas.

With purchases of nearby property, construction and renovation, the combined buildings of Sanger Brothers store eventually covered a downtown block between Main and Elm along Lamar. Alex Sanger handled the physical expansion and marketing aspects of the business, while Philip was the merchandising genius.

Chicago. A huge searchlight, the first of its kind in Texas, was used to illuminate this area of the grounds, and the spectacle of 12 cars revolving on a giant vertical wheel could be seen long before entering the gates.

Governor Hogg borrowed another gimmick from the Columbian Exposition when he opened the fair by touching a button to start the machinery. Visitors wasted no time filling a small theater for the first local performance by Turkish dancers. They walked the Streets of Cairo, mingled with the South Sea Islanders, wandered through the Irish and German villages and paid for a look at the "Wild Girl from the World's Largest Island," an Aborigine named Minnie who came from Australia. Some of the sights in this archeological zoo were free. Off-duty midway performers roamed the grounds riding on camels or being carried in sedan chairs.

The sun that smiled on opening day continued, and the numbers grew. Management moved Mrs. General Tom Thumb and her Royal Little People into the center gallery of the Main Hall to accommodate the crowds. Thousands were lured into Professor Roltair's Mirror Maze. A white-frocked maiden played the piano at the center of this glacial labyrinth, and the professor offered a $100 prize to anyone who could master the maze and locate her.

The birth of a baby in the Dalhomey Village during the first week of the fair caused considerable excitement. Dalhomey had been a center of African slave trade, and the infant was

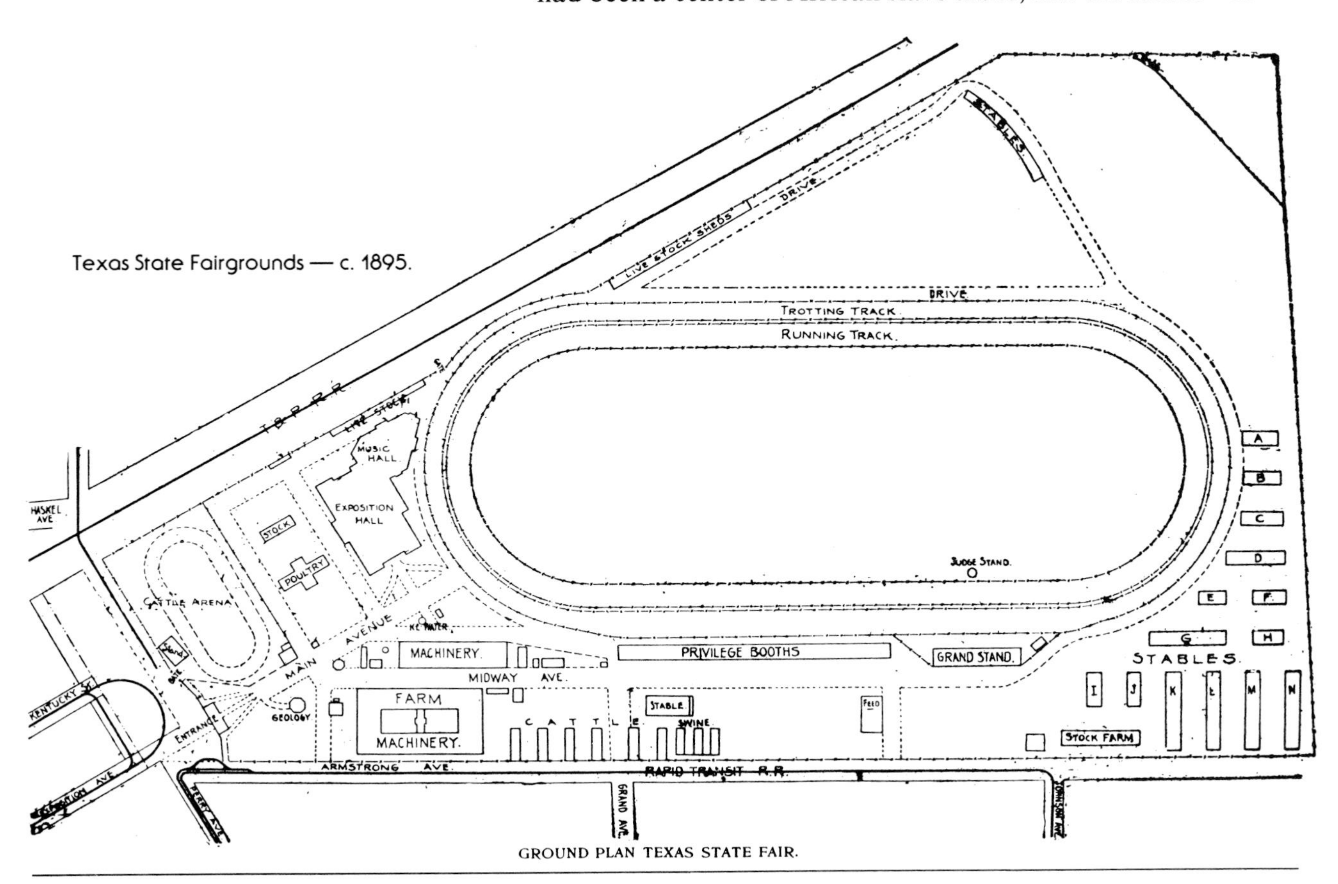

Texas State Fairgrounds — c. 1895.

hailed as the first real cannibal to be born in the United States. The tribe responded by naming the child "Texas" and complied with local laws that prohibited using human fluids for nourishment. "Texas" thrived on a substitute diet of chicken blood.

Fairgoers enjoyed exhibitions by Dr. W.F. Carver, the world champion rifle shot, and demonstrations of Thomas Edison's Kinetoscope, a machine for visually reproducing motion that the inventor described as a preliminary step in perfecting the moving picture camera.

Rain fell lightly and briefly on two overnight occasions, just enough to settle the dust. A few individuals complained about suggestive performances during midway shows, and police arrested numerous pickers and fakirs, the itinerant con men and pickpockets that attached themselves to major entertainment events, but problems were generally as mild as the weather.

For whatever his responsibility in maintaining sunshine for 16 successive days, the fair's president earned the nickname "Alexander the Great," and the phenomenon was popularly referred to as "Sanger Weather."

Well-accustomed to success, Alex Sanger ignored the accolades and kept busy with operational details, to the extent of working the front gate ticket window when necessary. When the window finally closed, the books showed a record $109,183 in receipts, enabling the fair to pay bond interest and other portions of its outstanding debts. President Sanger recommended a crackdown on complimentary admissions and suggested selling privileges on a flat fee rather than a percentage basis. The directors voted Guy Sumpter into the presidency, but Sumpter immediately stepped aside, and John T. Trezevant was elected to lead the organization in 1895.

John T. Trezevant, State Fair President, 1895-1896.

1895

Trezevant, a powerful insurance executive and financier, had a hand in many civic enterprises and a loud voice when it came to discussing the city's future — a future, as he reminded everyone, which would require facilities and services for a population of a quarter million rather than the present 50,000.

Under Trezevant's direction, the fair settled up four years of overdue taxes and embarked on a program of physical improvements which included a 1,000-coop addition to the Poultry Building, new roofs for the Machinery and Power Halls, flooring for the large open court in the center of the

Machinery Hall and construction of an 800′ restaurant pavilion near the grandstand to be known as privilege row. At the upper end of this row, an attractive little cottage was built to serve as a public house of comfort, especially for the ladies.

For 1895, the directors booked the ultimate name entertainer of the day, John Philip Sousa and his 50-member concert band. Trezevant commented, "Experience has taught management that visitors to the State Fair — the majority of them — attend more to seek amusement and be entertained, than as a matter of education, and while we propose to preserve the standards of the various departments . . . we will in a particular manner look after the attraction program."

This analysis of fairgoer's taste proved unfortunately accurate in the case of Sousa who exhausted most of the audience for his classical repertoire in the first week and played to many empty seats thereafter.

Challenger Bob Fitzsimmons

A heavyweight championship fight between titleholder James J. Corbett and challenger Bob Fitzsimmons should have been the main event of the 1895 fair.

Dallas sportsman Dan Stuart outbid all rivals for the rights to this long-anticipated match-up. The date was set for October 31. Stuart and a coterie of local businessmen, including several members of the exposition board, proposed to build a gargantuan amphitheater on property a few hundred yards northeast of the fairgrounds. The auditorium was designed for 52,000 spectators, making it, in terms of seating capacity and in terms that delighted its promoters, "the second-largest building in the history of the world."

City officials and railroad kingpins were ecstatic; religious leaders were outraged and opposed; and neither the governor nor the attorney general seemed sure whether prizefighting was even legal in Texas.

Stacks of lumber, nails and hardware awaited a ruling, while ministers preached and congregations prayed for divine intervention. In mid-September, a test case invalidated the law against pugilism, and construction began. It halted abruptly when Governor Charles Culberson, with an ear to the political winds, suggested that a new law could be written and called a special session of the legislature to make certain that happened.

The skeletal arena was torn down, and the championship rescheduled for Hot Springs, Arkansas, then Juarez, Mexico, and finally Carson City, Nevada, a nomadic progression that suited Corbett who hadn't been anxious to fight in the first place.

Eighteen months later, on St. Patrick's Day, Fitzsimmons knocked out Gentlemen Jim with a lethal blow to the solar plexus. The flickering filmed version of the bout was a big hit at the 1897 Texas State Fair and Dallas Exposition.

Among the less highbrow attractions offered were "The Last Days of Pompeii" — a 90-minute fireworks extravaganza; a snake show boasting 1,000 reptiles and advertised as "a mile of snakes"; a local terrier who had adopted and nursed an orphan kitten; and an oldtime corn shucking contest that was expected to draw elderly Negroes with first-hand experience in the art.

Participation in the livestock shows reached new highs. Trezevant, noting the Jersey, Holstein, Devon and Durham entries, pointed out the effect the fair was having on the cattle industry: "Texas is being rapidly 'dehorned' of her antlered cattle that had to live upon the prairies because their horns would not permit them to go through the woods."

Governor Culberson, whose hometown popularity had been polarized by his role in the boxing squabble, showed up for the opening, but the fair's high-profile speaker was the fiery young orator from Nebraska, William Jennings Bryan, who delivered the principal address on Free Silver Day. Days were also allocated to the Prohibitionists, Populists, Republicans and Sound Money advocates.

The ladies' department offered a new area where art and handwork could be sold and provided space to showcase items crafted by patients at the North Texas Hospital for the Insane.

Sparks from the fireworks show damaged a privilege tent, resulting in a small lawsuit against the fair, and a larger suit was filed by a man claiming his horse had been injured in a race. But 1895's most distressing event was a fire that swept through and destroyed the new row of restaurants shortly after the fair closed.

The exposition showed a profit, but as before, interest payments and past obligations made it impossible to significantly reduce the total debt which, including the bonds, remained well over $100,000. Suggestions that the

William Jennings Bryan — 1895.

organization plan a semi-centennial to mark the 50th anniversary of Texas statehood or consider building a new exhibit hall were countered by the imperative need to wipe out at least the floating portion of the debt, a responsibility which many directors felt should be assumed by the Dallas businesses enjoying the economic benefits of the fair. With this as a goal, Trezevant agreed to serve another term.

1896

The amphitheater debacle did not discourage sports entrepreneurs from selecting new sites near the fair. Baseball and racing fans already used the excellent public transportation system to attend events on the grounds, so it was natural for other promoters to locate close by. In 1896, two English-born businessmen roughed out a six-hole course on property along the northeast side and introduced the game of golf to Dallas, while adjacent to the southwest corner, C.R. McAdams built Cycle Park which featured a steeply banked wooden track for bicycle racing.

Construction was taking place on the fairgrounds as usual, but with a watchful eye for costs. Sydney Smith, who had relinquished his office as secretary to C.A. Cour, returned in a managerial capacity to supervise off-season work. Restaurant row was rebuilt just opposite the site of its predecessor, and the livestock department, which had been shifted three times and worn out its welcome with three sets of neighbors, was moved back to its 1886 location.

In an effort to create an exciting concert agenda for the Music Hall, Smith journeyed to Mexico City to request that President Diaz allow a musical group to perform in Dallas.

Restaurant row, sometimes called "Smoky Row" from the fumes and soot produced as churches and charitable organizations prepared hot meals for hungry patrons.

Diaz agreed to send a representative without cost to the fair and chose his wife's favorite, the First Artillery Band.

Novel attractions, such as a traveling Indian baseball team and a marine exhibit featuring a live whale and killer sharks, were booked, but it was impossible for management to come up with entertainment as bizarre as an event staged about 20 miles to the Dallas side of Waco three weeks before the fair. Passenger agent W.G. Crush sold tickets to watch two locomotives, each pulling six stock cars, collide at 60 mph. Crush's Crash attracted 20,000 spectators. At least one person was killed and several more injured.

The nearest comparable spectacle the fair could produce was a new racetrack feature pitting a Panhandle jackrabbit, with a 100-yard headstart, against a pack of greyhounds. The hare-hound events were planned for Sundays only, when horse racing was not permitted for moral reasons. The rabbits were trained and experienced, and as one newspaper pointed out, obviously good or they wouldn't still be alive. Disappointment was widespread when the first Sunday's races had to be cancelled because someone poisoned all the rabbits during the night. More contestants were shipped from West Texas, and an armed guard stationed at their improvised warren. The second program ran according to schedule. There were no long-eared survivors.

Season passes sold for $4.00, a lesser bargain than the $3.00 charge for an exhibitor's ticket. This disparity forced the ladies' department superintendent to issue a reminder that any exhibit entered had to be worthwhile and not just a means to qualify for the lower admission price.

Visitors were afforded the opportunity to watch x-ray demonstrations and showings of Edison's animated films, but sights and sounds took a backseat to the clouds and cold that bedeviled the 1896 exposition. Once again management was petitioned to extend the fair by three days, and this time the petitioners were businessmen who promised to declare a city-wide holiday on the closing date. The directors reluctantly agreed, and a grand informal parade was promoted for that final morning. A handicap bicycle race was also set, and such leading citizens as Trezevant, Sanger and Connor entered this one-mile event that included a rest stop on the back stretch where a separate contest would determine which rider could consume the most beer and sandwiches in one minute.

The parade attracted 200 vehicles and 15,000 spectators in mist and drizzle; the bike race was run in steady rain. Wind blew one of the tall smokestacks off the engine house, and water leaked through the Main Hall roof onto the displays.

By the end of the day, there was little cheering. Blame for the dreary 19-day run was placed on the weather, continuing hard economic times and distractions created by the approaching presidential election. William Jennings Bryan, campaigning on the issue of free and unlimited coinage of silver, lost decisively to William McKinley, a skillful political practioner whose aspirations had been underwritten and managed by Ohio industralist, Mark Hanna. Dallas supported McKinley, voting Republican for the first time.

In early November, city leaders called a meeting to discuss the fair's future. Trezevant's report indicated a disturbing drop in attendance by local residents, ongoing debt and a disheartened board of directors. The implement dealers, picking an inopportune time to resurrect old grievances, complained they had received minimal benefits from the recent exposition and threatened to pull out. Someone suggested that no fair be held the following year, so that Dallas people might better appreciate its importance. Mayor Frank P. Holland proposed that the city council consider issuing bonds to buy the fairgrounds and convert the property into a park.

A committee was formed, but nothing substantive developed. The fair directors paid what obligations they could, elected L.M. Knepfly as president and pushed on into 1897.

Lawrence Knepfly, State Fair President, 1897.

1897

The new president was the second-named partner in Knepfly & Son, proprietors of a fine jewelry store on Main Street since 1876. Lawrence Knepfly grew up learning his father's technical and commercial skills. At age

learning his father's technical and commercial skills. At age 50, he had established his reputation as a tough, precise businessman of unquestioned integrity. Although hampered by poor health, Knepfly and Sydney Smith, who was once again serving as secretary, set out to bring order and economy to the fair's day-to-day operation. The Main Exposition Building was finally reshingled and generally overhauled, and the grandstand in the ballpark was moved to the side of the field adjacent to the T&P tracks and essentially rebuilt.

It was a year characterized by extremes around the world. Cuban unrest brought the United States and Spain to the edge of conflict. Blight wiped out the Irish potato crop and produced devastating famine, while gold fever spread following word of a fabulous strike in the Klondike.

In September, a more serious epidemic concerned Dallas. Yellow fever raged in New Orleans with outbreaks also reported in Mobile, Memphis and parts of Mississippi. The death toll spiraled, and when cases were confirmed in Galveston and Houston, the Dallas Board of Health slapped a quarantine on residents of both cities. Even mail originating in New Orleans was put through a disinfecting process.

Knepfly vehemently denied rumors of a fair postponement. He obtained a $5,000 donation from railroad officials, and the exposition received strong support from the Dallas Manufacturers' Association which contracted for much of the space in the Main Hall.

The show opened as scheduled in mid-October with such features as an 80,000-pound whale, preserved with barrels of embalming fluid, and a replica of a prehistoric "ship lizard." This reptile was thought by some to be the ancestor of all mammals as well as a model for all sailing vessels. It measured 20′ long, carried five sails on its back and was burdened with a head too horrible to describe in daily newspapers. The professor in charge of the state geological exhibit speculated the lizard had become extinct because it could not adjust to a cataclysmic climate change, the same kind of change astronomers thought imminent if the sun, as expected, threw off another planet for the first time in 23 million years. Bible-minded speculators predicted this disturbance would usher in the millenium.

Curiosities continued to be popular at the fair. Visitors met the largest married couple in the world, Mr. and Mrs. Chauncy Moreland, whose combined weight totaled 1,300 pounds. They stopped by the stall of the oldest living Confederate war horse, examined the world's smallest operational railroad. Extravagant claims were seldom challenged. In the experience of most fairgoers, these attractions were indeed the biggest, smallest, oldest or only ones they had ever seen.

A new way of seeing was being introduced, however. Under various trade names — Cinematograph, Phantoscope, Vitascope — machines were being used to photographically recreate actions and expressions of persons and events far removed by time and distance. The prime example of this at the 1897 fair was the Verascope presentation of the Corbett-Fitzsimmons fight denied to the Texas public two years earlier.

Despite a slow start due to the yellow fever scare, the usual siege of wet weather and an entertainment lineup that was generally unspectacular, the 12th annual Texas State Fair surprised everyone by paying its expenses and some old debts as well. Knepfly accepted commendations, but reminded the directors and stockholders that under the present interest arrangement there was no hope of paying off the bonds.

Americans have a natural curiosity about who is building a better mousetrap, and mousetrap companies have a need to meet loyal customers and prospective clients face-to-face. This is the basis for the sale of exhibit space at the State Fair.

The fair has always been a unique advertising vehicle. In 1898, the directors decided to trade on their own marketing philosophy and deliver the exposition's message in a personal fashion.

A special railroad car traveled throughout Texas, New Mexico, Arizona and Louisiana, carrying posters, placards and handbills heralding the wonders of the Texas State Fair and Dallas Exposition. This flashy piece of rolling stock attracted thousands with its colored lights, Biograph motion picture equipment and Gramaphone talking machine.

The first venture was so well received that later trains spent up to four months distributing materials in outlying areas.

1898

The man most responsible for the fair's establishment and survival became its next president. Throughout his 12 years on the board of directors, W.H. Gaston had answered every call. He had loaned, guaranteed, endorsed, subscribed, waived, forgiven and donated. The association still owed him $40,000 in principle and accumulated interest from an 1887 property transaction, but Gaston's unwavering commitment encouraged support from others within the city and around the state.

On February 15, 1898, the smoldering situation in Cuba erupted after the sinking of the USS Maine in Havana harbor. America entered into what one diplomat called "a splendid little war" with Spain, an unnecessary confrontation between unequal forces. It lasted 113 days and spawned an aggressive, exhilarating patriotism across the nation. The United States had become a world power with possessions and long range interests in the Caribbean and Pacific.

This ebullient mood prevailed as the Texas State Fair organized its fall attractions. Peace Day was scheduled, and those not already sated by sensationalistic newspaper coverage from the front could view the action through Edison's latest moving picture device, the "War-Scope." A marionette troupe planned to recreate the "Battle of Manilla" for Music Hall audiences, and the featured outdoor showpiece was "Dewey's Victory in Manilla," a spectacle encompassing 900′ of scenery, 200 performers, war vessels and fireworks in the ballpark every night.

The entertainment lineup offered the Boston Ladies' Military Band, Nellie Chandler's Lady Orchestra from Chicago and the cornet-playing Mullins Sisters. Comic relief was promised by an act entitled "Heap Fun Laundry," which revolved around two Chinamen under the influence of opium chasing each other through a trick laundry.

The main entry gate in 1898.

Hoping to avoid rain, exposition management decided to open two weeks earlier than in the past and was rewarded by generally superb weather, even a few days described as "scorchers."

Enthusiastic crowds loved it all. They applauded the diver who plunged from an 80′ pole into 30″ of water, and they showed up to watch an ostrich being plucked. The gangly birds had been raised by a farmer in San Antonio who conceived the project as a crop diversification experiment. He argued that $45 ostrich plumes had more potential than four-cent cotton.

It had been a good production year for Texas agriculture, and the horticulture department showcased watermelons as big as beer barrels and pears the size of a man's head.

A first for the fair and a first for the state was the four-day show conducted under the auspices of the American Kennel Club. The competition drew 250 entries, 65 from Texas, to an area of the Machinery Hall outfitted with a canine kitchen and a huge bath kettle.

Special interest speakers took turns at the Music Hall podium. Barbers Day was highlighted by an address on "Barberism," and a congress sponsored by the Women's Christian Temperance Union considered the topic: "Should children be fed when they are hungry or at regular times?" This serious oratory provoked complaints every time it interrupted the regularly scheduled free entertainment.

It was a remarkably successful fair. The officers were re-elected unanimously, and the directors were able to take care of expenses, interest, past obligations and even pay Gaston $11,000 of the longstanding real estate debt. Only the awareness that time was running out for payment of the bonds shadowed the year's achievements.

Weather had presented no problems, but as Sydney Smith noted, the financial outcome of each fair hinged upon having 12-14 days without rain. In his mind, this was too slight a margin for a 16-day run, so he proposed going to a three-week format in 1899. Smith theorized that the additional per diem expenses could be covered by increasing the privilege fees, leaving the final week's receipts as clear profit.

1899

The association's venerable secretary got his request — a 25-day fair and approval for a long list of construction projects he had recommended. Before the year was out, Smith likely remembered the warning of contemporary playwright Oscar Wilde: "When the gods wish to punish us they answer our prayers."

Texas was producing one-third of the nation's cotton, and by 1899, Dallas had become the center of the cotton gin industry. The city's two large factories, the Munger and Murray companies, and six others from across the United States offered to assemble a complete gin exhibit if the fair would erect a building for the display. To accomplish this, Billy Gaston bought 100 carloads of lumber on credit and got them shipped from Orange to Dallas at half the usual rail rate, a feat that infuriated local lumber merchants and resulted in Gaston and Smith being called before the Texas Railroad Commission. Their permits were in order, and work began on the Gin Building, an imposing structure powered by 300-horsepower boilers and engines. A 65′ well was sunk nearby to provide a non-alkaline water source for the steam operation.

The old Machinery Building was renovated to create a 3,000-seat auditorium and a spacious flower hall. Additional seating and new curtains were installed in the Music Hall, and an exhibition kennel was built for the dog show. The ladies of the WCTU erected a small rest cottage for the day nursery they sponsored each year.

Billy Gaston's responsibilities as president of the fair occupied him through the difficult period following the death of his wife, Ione, in June of 1899.

The new auditorium was selected as the site for the first Kaliph's Ball. Under the direction of liquor wholesaler Charles

The Grand Order of Kaliphs was organized to produce a city-wide celebration comparable to New Orleans' Mardi Gras.

Mangold, the Grand Order of Kaliphs had been organized to present a gala civic entertainment along Mardi Gras lines. The three-day program would take place during the fair and include a welcome for the ceremonial Kaliph of Bagdad, a nighttime parade of illuminated floats through downtown and a grand ball that promised to rival any existing social event in the state.

The National Democratic Carnival, a pre-election year conclave for William Jennings Bryan and other party leaders, was also scheduled in Dallas for the first week in October, and fair officials expected these added activities would bring huge crowds to the grounds.

As it turned out, the Kaliphs and Democrats stole some of the exposition's normal thunder, and attendance lagged badly through the first ten days. Then the drought that had been causing water shortages in North Texas ended. When the rain let up, management ran ads assuring everyone that the best of the fair was still to come. Gaston arranged for a repeat performance of the Kaliph's Pageant under hastily-strung lights above the racetrack and prevailed upon the railroads to cut rates again. These measures helped, and thousands turned out on the final Saturday to attend the first-ever reunion of University of Texas alumni and friends. A football game followed the speeches. The Texas team had played in Dallas and at the fairgrounds before, but never on an occasion where their rooters outnumbered the opponent's. The UT fans pushed, howled and generally misbehaved. One spectator clubbed a Dallas player with a cane after he was run out of bounds. Providentially, darkness halted both the action on the field and an incipient riot in the stands.

At the close of the 24-day run, the Manchester Trust Company demanded full payment on the bonds it held. When the fair was unable to comply, the English company brought suit, secured a judgment and advertised the property for sale in an effort to satisfy its claim.

To save the fair, the directors petitioned for a receivership. This stay of sale brought about a conference of all the bondholders which produced an agreement to foreclose, sell, reorganize and rebond under more favorable terms. Gaston accepted $6,000 as final settlement of the money owed him, and a new corporation was formed.

1900

There was disagreement whether the twentieth century started at the beginning or end of the year 1900, but harbingers of the future were already on Dallas streets.

Ned Green, railroad executive and son of the notoriously wealthy Hetty Green, drove his two-cylinder, surrey-styled automobile into town on October 5, 1899. The new horseless carriages cost about $1,500, well beyond the reach of the average worker who earned $12 weekly, but the vehicles multiplied, ushering in an era of asphalt paving, traffic ordinances and automobile dealerships.

William McKinley campaigned for a second presidential term by promising "four years more of the full dinner pail." Most Americans believed in the inevitability of progress, although people in Dallas were chagrined to learn the latest population census dropped them into third place behind San Antonio and Houston.

President Gaston and the fair directors expressed renewed optimism. They borrowed money to patch the Exposition Hall roof again and applied liberal coats of paint wherever it was needed.

In early September, a hurricane struck the Texas coast virtually destroying the island city of Galveston. The disaster, which took 6,000 lives and left thousands more homeless, touched nearly every community in Texas with the loss of relatives and friends.

When the fair opened on September 29, the grounds were so filled with tents, booths and activity centers that a "parking" crisis developed. Management announced that a portion of nearby Cycle Park would operate as the official wagon yard and vehicle repository. Horses would be cared for — and fed, if desired — at reasonable prices.

In 1900, buggy exhibits competed for fairgoers' attention, but by 1904, some of these were being replaced by displays of "horseless carriages."

Racing continued to be the fair's most popular attraction and primary source of income.

A new sport was introduced to fair visitors. Polo matches were played on the racetrack infield. To assure an audience, badger fights and wolf chases were staged in conjunction with the equestrian event. Members of the polo teams used lariats to rope any wolf that escaped the hounds.

For the first time, the ladies' department offered a miscellaneous class which gentlemen were permitted to enter. Family relics, old books, documents and curios were judged in this category. The adjacent art department showcased a loan collection of paintings by the acclaimed frontier artist, Frank Reaugh, an Oak Cliff resident.

Belgian hares, a fad said to be sweeping the country, caught the attention of fairgoers. More than 500 of the unusual creatures were exhibited by a California rabbitry. The poultry department also featured a pair of Pelamopedogosanders, rare South African geese with red plumage.

The fair's big entertainment spectacle was "The Burning of Chicago," a production depicting all the horror of that 1871 conflagration in a reconstructed city block. This mammoth show, which included real fire engines, burning buildings and heroic rescues, all displayed in a dazzling shower of fireworks, was set up in the old baseball park.

A second Kaliph's Carnival was scheduled, and workmen transformed the auditorium into an appropriate setting for a grand ball. On that Wednesday evening, as Dallas society enjoyed dancing and refreshments, another crowd of 3,000 had gathered to watch the fireworks show. Sometime after 8:30 p.m., a section of the old ballpark grandstand collapsed. Spectators in the top rows tumbled down upon those in the lower rows. Some were partially buried under broken timbers, but miraculously only one person was hospitalized in serious condition.

Good weather and good crowds put smiles on the directors' faces. Attendance was estimated between 50,000-60,000 on the middle Sunday. Hotels and boarding houses had to turn people away that night, and those who couldn't squeeze onto outbound trains slept on depot platforms. Another huge crowd arrived on Colored People's Day to hear the illustrious Negro educator, Booker T. Washington.

But the new one-day record was established on Buffalo Bill Day. Approximately 70,000 people packed the fairgrounds to see the exposition and two special performances of Colonel William Cody's extravaganza. Cody had toured with smaller companies, but this was the "big show" of World's Fair fame. It starred Miss Annie Oakley, 600 horses and a buffalo herd.

Booker T. Washington delivered the principal address on Colored People's Day in 1900.

Sydney Smith's report showed that the 1900 exposition ranked as the most successful ever, with enough profit to make a substantial payment on the new bonds. Within 60 days, however, damage suits to the amount of $150,000 were brought against the fair on behalf of persons hurt in the grandstand accident. Most of these claims were based on internal injuries to brain tissue, sciatic nerves or female organs. Memory loss was also cited, and one man insisted the fall had shortened his leg. Legal fees and court costs became another major expense on the books of the 15-year-old organization.

1901

In 1901, President McKinley, who had won a lop-sided second victory over Bryan, spoke confidently of prosperity and peace, and a gusher that blew in on a salt dome south of Beaumont guaranteed phenomenal economic growth in Texas. The Spindletop field would produce 50 million barrels of oil in the next decade. Other discoveries, refineries and industries spawned by the demand for drilling equipment and pipeline created vast wealth for individuals, companies and the state itself.

As the year began, an era ended with the death of England's Queen Victoria. That spring, the Pan American Exposition opened in Buffalo, New York. Designed "to promote commercial and social interests among the states and countries of the western hemisphere," this small-scaled international show literally spotlighted Niagara Falls to demonstrate electrical power and introduced the wireless telegraph and Edison's new storage battery to an eager public.

Texas State Fair directors were interested in ideas and attractions that might be brought to Dallas. Their attention turned to Buffalo in early September when President

Livestock facilities, including this show pavilion, were rebuilt north of the racetrack in 1901.

McKinley visited the Pan American to deliver a speech during which he stated that "isolation is no longer possible or desirable." The next day, while attending a reception at the exposition's Temple of Music, he was wounded twice by an assassin and rushed to an emergency hospital inside the park.

McKinley died eight days later, and in the aftermath, fair officials in Texas announced that an emergency medical tent, patterned on the Buffalo facility, would be established on their grounds.

The unsightly row of sheds that served as the livestock department was moved again, this time to the north side of the racetrack east of the Music Hall. Roadster horse racing was added to the agenda, and the Mexican Village boasted a new option for fairgoers — a Mexican restaurant.

All State Fair advertising carried a disclaimer that the association assumed no responsibility for "injuries, damages or losses to any article, animal, fowl or person," but suits from the previous year's accident remained unsettled.

Auto racing made its State Fair debut. Sunday programs featured matches between two full-sized cars plus some motorcycle vs. horse contests. Incubators and brooders, nicknamed "wooden hens," were exhibited for the first time in the poultry department.

Attendance was good, though not up to the 1900 exposition's standard, and the re-elected fair officials prepared for a busy off-season that began with the biggest convention Dallas had ever hosted, the 12th National Confederate Soldiers Reunion.

C.C. Slaughter, one of Texas' best known cattlemen, was chosen as commanding officer for the reunion. Over 25,000 people were expected to attend the four-day event in April. Old soldiers who chose not to stay in hotels or private homes would be housed in camps and tents on the fairgrounds and fed daily without charge at 700′-long tables. Colonel Slaughter brought five buffalo from his ranch to be butchered and cooked for this feast. All formal meetings and entertainment, including a concert by piano virtuoso, Ignace Paderewski, were scheduled in fair facilities, and the auditorium was overhauled to accommodate these activities. During the remodeling, part of the ceiling caved in killing one carpenter and injuring 15 others. The reunion was a grand success, but the fair association was targeted with another $60,000 in damage suits.

The worst was still to come. Early on Sunday morning, July 20, 1902, fire broke out in the Main Exposition Building and within an hour completely destroyed that structure, the adjacent Music Hall, the Poultry Building and three small vehicle exhibit quarters. The blaze could be seen as far west as Fort Worth and as far east as Wills Point. Intensive effort by three Dallas fire companies saved the rest of the grounds, but by morning, blackened, twisted iron and charred shrubbery was all that stood on the Exposition Hall site.

There was neither time nor money to rebuild, so the directors worked with what remained and relocated various departments. There were fewer chickens and no jams or jellies, but the exposition opened on time in a modified setting. President Gaston emphasized that the prospects of replacing the losses would depend on the success of the current fair, a remark directed particulary at Dallasites who had been increasingly less supportive of their annual day.

The Universities of Texas and Oklahoma played football at the 1902 State Fair — but not against each other. Oklahoma lost to an all-star team from Dallas, while Texas triumphed over Sewanee.

Entertainment was provided by an aquatic act that featured trapeze stunts, dives and speed swimming in a 20′ x 50′ tank. The Olympia Opera Company presented a series of comic operettas, and the Victoria Electric Fountain, a portable spectacle of water and lights, was set up on the now-vacant main plaza for evening shows.

The usual contingent of politicians visited throughout the run, but the special guest who attracted the most notice was Carrie Nation. The "Kansas Saloon Smasher" had been invited by the WCTU to speak on Prohibition Day, and a curious crowd tagged along during her official inspection of the grounds. She stopped briefly in the ladies' department to comment, "This fancy needlework doesn't suit me. Women spend a whole lot of valuable time on it, and it doesn't amount to anything after it is finished." The tour concluded, at her

Downtown parade — 1902.

request, in front of a privilege bar near the auditorium. Miss Nation berated the bartender, dumped one customer's whiskey on the ground and marched inside to deliver her address.

The fair made a profit, but not enough in view of its staggering needs and obligations. Again there was talk of the city taking over the grounds and paying off the bonds. Committees were appointed, followed by discussion, but no action was taken.

The proverbial back-breaking straw landed squarely on the association in 1903. The 28th Texas Legislature, responding to complaints about gambling abuses, banned pool-selling and betting on horse races.

1903

Racing had always been the State Fair's main attraction and main source of revenue. With little confidence or choice, the directors substituted a long list of grandstand acts. Horse racing without betting was financially impractical, but spectators would be able to watch bronc busting, a hurdle competition for mules and exhibitions between running horses and Black Diamond, the racing ostrich.

Thick foliage covered many areas of the grounds. The original Machinery Hall, later designated as the Manufacturers Hall, then renovated and renamed the Auditorium in 1899, was used as the primary exhibit building after fire destroyed the Exposition Hall in 1902.

A landscape gardener was hired to restore the burnt-out areas with grassy plots, flower beds and tropical plants. Creeping cyprus vines trailed over buildings, arbors, fences and trees, effectively doing the work of a paint brush in hiding defects.

A mammoth white tent was erected on the old Exposition Hall site to present Kilpatrick's Loop-the-Loop Show, an expensive, much-heralded attraction. Dozens of galvanized iron ticket boxes were taken out of storage, and the 1903 Texas State Fair, eighteenth and last of its kind, started admitting visitors on September 26. One familiar face was absent. Margaret Smith, wife of Sydney and the organizational force behind the ladies' and art departments, died shortly before the opening.

Kilpatrick was a one-legged cyclist who forged his reputation by riding a two-wheeler down the steps of the Capitol Building in Washington, D.C. He performed an approximation of this at the fair, but the star of the show was Diavolo, an intrepid rider clad in a Satanic costume, who sped down a 100′ incline into a concrete loop where he negotiated the complete 360-degree circle without falling off the bicycle onto his head. The difficulty of this stunt could be gauged by Kilpatrick's standing offer of $200 to any amateur who could duplicate it. Of those who had accepted this challenge in other

Fairgoers waiting for transportation outside the secretary's office at the front gate in 1903.

cities, 40 had tried and 40 failed. Two were killed, and no one escaped without broken bones. The show's other headliner was Kiro II, a monkey who drove a tiny automobile through a smaller loop. His predecessor, Kiro I, reportedly had gone insane the week before the Dallas engagement.

The fair was an honorable attempt in the face of overwhelming odds. It rained the first Monday, and Billy Gaston said cheerfully, "We cannot expect the sun to shine every day . . . and if we were to have rain, this is the very day we'd have picked for it." When the next storm blew down part of the Kilpatrick tent, the fair's president had no comment.

In the final analysis, no one blamed the weather. Total receipts had dropped nearly 50%, and one official observed that a fair without racing was like "Hamlet" minus the melancoly Dane. A $2 premium award mailed to a Dallas widow was returned with a note saying the fair needed money worse than she did.

Developers had been interested in the 117-acre fairgrounds for several years, and very quickly the directors received a cash bid of $125,000 from a St. Louis syndicate that wanted to convert the property into a suburban addition. This amount would have cleared all indebtedness and permitted a sizable reimbursement for the existing stockholders.

It was a tempting proposal, but acceptance meant that Dallas would no longer host the State Fair. The stockholders and directors had invested too much in this enterprise to see another Texas city reap the benefits. They rejected the offer and prepared a prospectus, largely based on a city/fair partnership that had worked successfully in Toronto for 20 years, which they presented to the Dallas City Council in February. Under this plan, as modified by the council, the City of Dallas would pay $125,000 for the real estate and all other association assets. Funds for the purchase would be generated by a special property tax to be levied over a four-year period. The directors would use $80,000 of the sale price to liquidate the bonds. The remaining $45,000, plus an additional $30,000 to be raised from the sale of stock, would be donated to the city to construct a modern, fire-proof Exposition Building. The terms further stipulated that while the grounds and improvements would belong to the city to operate as a year-round public pleasure park, the fair association would be permitted to use the site for its annual exposition.

On April 5, 1904, Dallas voters approved this proposal by a 2,531 to 415 margin.

The fair was reorganized and renamed the State Fair of Texas. New stock was issued that specified no dividends would ever be paid, and the bylaws were rewritten to set up a non-profit corporation. The fair and city signed a contract which, with only minor changes, still governs the park's operation. The State Fair of Texas agreed to produce the event of that name, pay operational expenses and reinvest net revenues from the annual fair in maintenance, park improvements or new attractions for the people of Dallas.

1904

The mechanics of restructuring the fair and transferring the property to the city extended through the summer into the fall. With no Exposition Building, no betting revenues and the reasonable expectation that their efforts would be overshadowed by the Louisiana Purchase Exposition, the World's Fair taking place in St. Louis, the new board of directors decided to present an abbreviated fall carnival which they named the Texas Grand Festival and Kaliph's Celebration. No premium book was issued, since at this point no money was available for awards. The nine-day event offered a mix of amusement attractions, fireworks shows, automobile exhibitions and horse racing without gambling. Other activities were postponed until 1905.

The anticipated competition from St. Louis never materialized. The colossal World's Fair spread over two square miles and offered 15 important exhibit palaces among more than 1,500 constructions. Its size was literally breathtaking and taxed the physical and mental capabilities of visitors.

State Fair officers in 1904: (from left) President C. A. Keating, Charles Mangold, Edwin J. Kiest and Sydney Smith.

This, combined with a hot, humid Missouri summer, kept attendance well below projections. Texans traveled north to see the show in its first few months, but interest dropped off considerably by fall. The St. Louis World's Fair introduced iced tea, ice cream cones and the popular song, "Meet Me in St. Louis, Louis," but it was a disappointment in scientific and cultural areas and closed with a substantial deficit.

One noteworthy feature at the St. Louis fair was an exhibition of 160 automobiles. Local fair officials arranged for a similiar, though smaller display of steam, electric and gas-powered vehicles on the track before races, in effect creating the first automobile show in Dallas.

There were changes in the leadership ranks of the newly-styled State Fair of Texas. Gaston remained a director and Sydney Smith continued as secretary, but the prominent figures in 1904 were Cecil A. Keating, president, and Charles Mangold, chairman of the attraction and amusement committee.

Keating, a leading implement dealer and sharp critic of fair management on past occasions, had been off the board of directors for ten years. During this time he had been the primary force behind the Trinity navigation project, an enterprise which Keating and others expected would be completed by government funding. Genial Charley Mangold had built his fortune in the wholesale liquor business and now managed Lake Cliff Park, a popular summer recreation center.

Construction prior to the festival was limited to a new band stand, another rest cottage and a few privilege booths. The focus in 1904 was the amusement area. Old "Smoky Row" had been brightened with gallons of white paint and renamed the "Trail," a fitting choice in a cattle-raising state.

The Trail featured a seated enclosure for circus and vaudeville acts, an outdoor dance pavilion and a burro ride concession. Mangold booked this entertainment and ran the five-day race meet. Both drew respectable crowds. The amusement features made money — horse racing, without betting, did not.

The festival failed to pay for itself, but provided important continuity. In December, plans were approved for a massive, classically-arched Exposition Hall for the State Fair of Texas, a building intended to last a century.

A turn-of-the-century amusement ride based on the principles of the giant wheel built by George W. Ferris for the 1893 Chicago World's Fair.

The Great State Fair

1905-1918

1905

Cecil A. Keating, State Fair President, 1904-1905.

A 100-mile automobile race around the fairgrounds' track on New Year's Day, 1905, marked the beginning of a new era in the history of Fair Park and the State Fair of Texas. The passing of the old was tragically symbolized a few weeks later when James B. Simpson, the association's first president, died in a nineteenth century kind of accident — he was thrown from a buggy while trying to halt a runaway horse.

With construction beginning on the Exposition Building plus park-related maintenance, rental requests and fee disputes to consider, the Dallas City Council was having difficulty getting through its weekly agenda. Seeking relief from these administrative duties, the council created Dallas' first park commission to oversee the city's two parks.

Legislative action in Austin that spring also had direct bearing on the fair. After months of debate, an amendment to the anti-gambling statute was passed which exempted

Texas governor S.W. Lanham and State Fair president C.A. Keating wait for the beginning of the 1905 parade.

The new Exposition Building, showing the ticket office and entry to a 3,500-seat auditorium that occupied approximately one-third of the available floor space.

wagering on the day and in the enclosure where a race was run. Fair officials were understandably jubilant. They built a new paddock for the track and prepared to offer the largest purses in the fall meet's history.

Local enthusiasm for the fair grew as the magnificent Exposition Hall started to take shape. Built of cement stone with three porticos sheltering the entrances that faced the main drive, the building contained more than 75,000 square feet of floor space. The doorway nearest the front gate led to a 3,500-seat auditorium. An interior brick wall separated this facility from the larger exhibit area under the second and third bays. The building ultimately cost $90,000.

Thousands attended dedication ceremonies on opening day. Fairgoers found the exposition rejuvenated and rehabilitated. The new plaza fountain, manicured lawns and emerald foliage drew appreciative comment, as did a collection of plaster statuary acquired from the St. Louis World's Fair. The statues had been erected throughout the grounds — some on appropriate sites, others in strangely incongruous settings. Towering mythological figures stood on top of the Exposition Building's outer walls, on the brick arch previously used as a post office, on top of the band shell and even on the roof of the women's rest cottage.

The Trail was now called the Pike, after the St. Louis amusement area of the same name. Two rides were introduced in 1905: the Figure Eight, a single track operation with rapidly descending dips and dives, and the mammoth Carry-Us-All, an improved merry-go-round with three rows of horses that galloped up and down instead of simply traveling in a stationary circle.

The most talked-about attraction from St. Louis proved equally fascinating to Texans. The Igorrote Village was (right) Members of a "headhunting" tribe from the Philippine island of Luzon inhabited the Igorrote Village in 1905.

(below) Plaster statues, acquired from the St. Louis World's Fair, were used for grounds decor, as here on top of the bandstand.

inhabited by 32 Philippine natives who went about their daily chores and entertained spectators with spear-throwing contests. They wore little or no clothing and, per tribal custom, they dined on dogs for their evening meal. Management was responsible for furnishing three large dogs every day. Presumably there was no connection between the Igorrotes' appearance at the fair and the decision by the Texas Kennel Club to discontinue its annual show.

It rained sporadically, but nothing discouraged the crowds. On Dallas Day, 65,000 paid admission and another 10,000 or so entered on passes to establish a new one-day high. Total attendance was estimated at 300,000. On the last night after the final show in the new auditorium, C.A. Keating took the stage to pronounce the fair "safely landed and at last beyond any danger of failing." Joined by the other directors, he led the audience in singing "Auld Lang Syne."

1906

The next State Fair president would take office with an advantage over his predecessors — $52,000 in working capital. The directors' problem was finding someone who wanted to be president. Keating firmly declined to serve another year. Sam Cochran also refused. Finally, they prevailed upon 55-year-old James Moroney to accept what had become a very time-consuming honor.

Moroney was a popular choice. The affable Irishman had been the association's second president almost 20 years earlier. His first recommendation to the board was that improved roads, drives and sewers be given high priority and that a top landscape architect be employed to study the grounds and plot future buildings.

The arched entry at Grand Avenue.

In March, the fair hired George Kessler of St. Louis to design a master plan for developing the property now commonly referred to as Fair Park. Kessler was particularly qualified for this assignment — he had grown up in Dallas. After studies in Europe, he established his reputation as a pioneer planner for the Kansas City park system and later as the designer of the St. Louis World's Fair site.

Kessler prepared a map which identified all structures and facilities he considered necessary for the fair's optimal development. In general, he endorsed the existing street layout and recently-completed Exposition Building. But he strongly recommended that an Administration Building immediately be erected at the main entrance.

The directors agreed. The old front pavilion, which by now consisted of an unsightly, weather-beaten line of sheds covered with torn posters and scaling paint, was leveled and replaced by a concrete stone office structure designed with three arches in its center to serve as pedestrian gates. The obsolete high board fence, rickety secretary's quarters and old Gin Building were also torn down. Arched entries were constructed along the southwest side at Grand and Trezevant avenues to relieve traffic congestion at the front gate. A new Poultry Building was erected, a new ceiling installed in the auditorium and paved sidewalks built along the major streets.

The Shoot-the-Chutes and Scenic Railway were constructed in 1906 as permanent rides for the park.

With no city funds available for expansion, the Park Board enlisted the aid of twelve businessmen who provided the cash under a purchase option agreement to buy a block-wide, 10-acre strip which extended the park's boundaries to Pennsylvania Avenue. This acquisition became necessary after the owners of these lots announced their intention to build homes with backyards opening into Fair Park.

Other entrepreneurs were looking for ways to make money using park assets. One firm obtained permission to convert the Machinery Hall into a skating rink during the off-season. Approval was also granted to build and operate two year-round amusement rides: the Scenic Railway and the Shoot-the-Chutes.

Fairgoers were delighted with the improvements. The Pike was spectacular. In addition to the permanent rides, a Double Whirl and Captive Balloon were new to North Texas. Amusement Row offered 110 different shows including a three-ring circus and baby incubators occupied by living babies.

For the more discriminating visitor, the art department featured 91 paintings, many on loan from collections in New York and Boston. Two renowned operatic soloists presented concerts during the fair. The entertainment by Signor Campanari and Madame Sembrich required separate tickets, priced as high as $1.50, and management protected the interests of performers and audience by refusing to admit small children.

John L. Sullivan, the last of the great bareknuckles boxers, headlined a one-night vaudeville special. The former champion's contribution to the show was a short monologue and an uninspired sparring exhibition.

The 1906 State Fair of Texas broke all the records established the year before. Single day attendance reached 100,000 on the middle Sunday, and total receipts topped the previous mark by $50,000.

1907

The annual report showed a cash balance of $85,000, enough for the re-elected president and directors to follow through with earlier plans to pave Fair Park streets with macadam and lay a system of sewer pipes. What remained of the old ballpark was torn down and the area replanted in lawn. Athletic events were shifted to nearby Gaston Park.

Year-round use of the grounds continued to increase. Movies were shown, and automobiles could be rented for

An exclusive new clubhouse for privileged racing fans.

pleasure drives through the park. One off-season enterprise did not appear to be doing well, however. The skating rink in the old Machinery Building needed maintenance, and the tenants had fallen behind on their rent. The Park Board tried to work out these problems with the operators until it learned that sleeping apartments had been constructed in this facility, and women had been observed "entertaining" gentlemen after closing hours.

Two Dallas institutions, destined for a profitable symbiotic relationship, made their debut in 1907. Herbert Marcus, together with his sister, Carrie, and her husband, Al Neiman, opened a specialty store downtown; and John S. Armstrong, who had acquired 1,326 acres along and north of Turtle Creek, turned this property over to his sons-in-law, Edgar Flippen and Hugh Prather, to develop a restricted residential community — Highland Park.

Fair Park added an exclusive amenity. A clubhouse for viewing the horse races away from common vulgarity and intrusion was built next to the grandstand.

Governor Thomas Campbell almost missed the fair's first day festivities: his train was late, breakfast service was slow, and the automobile carrying him to the park snapped an axle en route. After much delay, he delivered an address filled with platitudes and profundities to open an exposition which advertised itself, as it had the year before and would the year following: "bigger and better than ever."

As usual, the show lived up to its billing. For the first time, vehicles with gasoline engines outnumbered older-style farm machinery. A patent exhibit, showcasing the inventiveness of Texas tinkerers, featured everything from insect exterminators to re-designed clothes pins. An apiary and honey products display extolled the importance of another growing Texas industry.

Visitors cheered for "Joe Joker," the famous guideless pacer, who demonstrated a lot of horse sense and perfect racing form while circling the track without a driver. Long lines waited to see performances by a troupe of educated fleas, and the ostrich races delighted spectators. Texans loved these long-necked feathered aberrations. Their commercial value in providing gaudy plumes for women's hats was duly appreciated, but crowds came to watch the big birds get their legs tangled up in the harnesses, bury their heads in the sand and otherwise frustrate their drivers.

The Vice President of the United States, Charles Warren Fairbanks, spent an afternoon touring the exposition, and the Five Million Club was accorded a day to encourage immigration to Texas. This organization's stated population goal was five million by 1910, a sizable increase over the three million recorded in the 1900 census.

When the shouting died down, the dust settled and fair officials counted the take, there was a net profit of $93,000 to underwrite major improvements in 1908. For their new president, the directors elected Edwin J. Kiest, publisher of the *Daily Times Herald*.

1908

Kiest had purchased the floundering newspaper in 1896 and turned it into a profitable operation. He was the son of a Methodist minister, raised in small Illinois communities where his father held pastorates. Still a youthful 46, Kiest was prepared to provide vigorous leadership for the fair, beginning with the erection of a separate building for the ladies' and art departments, a project he had championed for years.

Drawing inspiration from the Columbian Exposition's "white city" architecture, the Textile and Fine Art Building was constructed of cement blocks and steel in a 125′ square crowned with a crystal-paned dome. Built for $37,000, it was located between the front fence and the auditorium end of the Exposition Hall in accord with Kessler's master plan. The majestic, if somewhat impractical dome, covered a central art salon. Mural cabinets for textile and curio displays formed walls in the outer corridor.

But the art facility was only one of three costly structures completed in 1908. The success of the racing program was underscored by construction of a modern stone and steel grandstand that was intended also to function as the city's major outdoor auditorium for important public speeches and convention activities. The grandstand, including its numerous comfort features, cost $44,000, and another $25,000 was spent on a spacious Agriculture Hall.

The Textile and Fine Art Building — 1908.

The Park Board hired James Flanders to build a vehicle gate at the front entrance, and the great masses of shrubbery and trailing vines were replaced by lawns and cement walks laid out in precise geometric patterns.

The bright outlook for the 1908 fair was clouded by a revival of public sentiment against bookmaking and pool selling. Persistent complaints of fraud convinced the board of directors to send representatives to investigate track operation in the north and east. Ben Cabell and M. M. Phinney reported that bookmakers were about to be driven away from every first class track in the country. In Kentucky, however, they had found an innovation called parimutuel betting which virtually eliminated gambling abuses.

The directors voted to install the new system, then discovered that existing statutes required that any reputable bookmaker willing to pay the fee be permitted to work the track. Rather than operate two incompatible systems, the board postponed adoption of parimutuel wagering until they could petition Texas legislators to rewrite the law.

A perceptibly-changed park greeted fairgoers in the fall. Display quality and presentation improved by virtue of the new facilities. A fine dairy exhibit attracted people to the center of the Agriculture Building. Texas dairymen, anxious to promote their infant industry, set up a working creamery where butter was made twice daily and displayed in a huge glass-paneled refrigerator. They pointed out that certain midwestern states, namely Iowa, Wisconsin and Minnesota, had grown rich from an industry for which Texas, with ten months of the year available for open feeding, was better suited. Nearby, a small broom factory turned out brooms, and cotton mill looms produced dishtowels.

In other buildings, visitors were alternately dazzled by Parisian gowns, with $2,000 price tags, and educated by a grim series of photos, charts and models telling the story of tuberculosis, which still ranked as the nation's number one killer. A light flashed every 2 minutes and 36 seconds as another American died from the disease.

At the 1908 fair, the "Sport of Kings" shared its spotlight, as well as the track, with a new sport of roaring engines and exploding tires. Owners of gasoline-fueled automobiles were less gracious and refused to race against steam-powered vehicles in a dispute over who could keep up with whom. A steamer settled that question, temporarily, by turning in the best time of the day in an exhibition run. With a crowd of 25,000 in the stands and lining the circuit, drivers pushed their cars to 60 mph speeds. The railbirds had to scramble for safety when one of the favorites, Fred Dundee, lost control on a curve, crashed through the fence and plunged into a tent full of agricultural products from the Panhandle. Aside from some torn canvas and damaged displays, there were no serious injuries and the race continued. For spectators, the dream of automobile ownership had become a viable possibility that year with the introduction of Henry Ford's first Model-T.

A bird's-eye view of the park showing the emergency medical tent and a booth sponsored by the Prohibition Party in the foreground.

Elsewhere, the sanctity of life was challenged by "The Great Velaire," who would mount his wheel, plummet down an incline and shoot out into space as the wheel dropped into a net. The airborne showman somersaulted twice and dove into a shallow tank located 65′ away from the base of the chute before taking a bow.

Racing fans responded to the splendid new grandstand by dressing up in fashions appropriate for a New York theater. Women wore gloves, long gowns of silk or satin in subdued fall colors and chic hats adorned with bird's wings or feathers. Even the maligned bookies toned down their customary behavior and attire. According to one report, they were a quiet, genteel group who shunned vulgar displays of diamonds and could have been mistaken for bank clerks.

Attendance for 16 days was estimated at 800,000, again providing the directors with nearly $100,000 for further park improvements. At the annual meeting, stockholders praised these accomplishments, then unexpectedly proposed that the exposition close its gates on Sundays in the future. The directors protested that this would eliminate revenue without decreasing expenses and would deprive many wage earners of their only opportunity to see the event. The resolution was tabled, but a breech developed between the fair and the Dallas church community.

1909

The breech became a battleline in the first days of the January legislative session when a bill was submitted making it a criminal offense to wager anything of value on a horse race anywhere in the State of Texas. The churches righteously supported this measure, branding the fair a "sink hole of iniquity" and accusing its directors of aiding and abetting the Devil.

Fair officials and horse breeders backed an alternative bill to abolish bookmaking and replace it with the parimutuel system, but they were outnumbered and out-lobbied by an unholy alliance of bookmakers, pool sellers and preachers. The anti-race bill passed, and the fair re-evaluated its expansion plans.

Money had been set aside to build a Coliseum for livestock judging and horse shows, but under the circumstances, President Kiest felt this should be used to increase purses and premiums to produce the best fair possible. To avoid a construction delay, backers of the Coliseum project offered to raise the necessary amount privately and loan it to the fair. The board members approved this idea, but quickly learned that the loan had strings attached; those who were furnishing the funds wanted a say in where the building would be located. Their choice was a prime site just inside the main gate. From earlier experience the directors knew this would never work. It would be impossible to lead bulls and stallions from the barns through fairtime crowds to the Coliseum, and it would be intolerable to stable the animals anywhere near the Art and Exposition buildings.

In a compromise, the new facility was constructed on the coveted location at the front of the grounds, but judging activities were retained in the barn area. The handsome brick and stone structure would be reserved for four evening horse shows and converted into a Music Hall for the remainder of each fair. With modifications for seats, staging and people comforts, costs escalated to $108,000, but the building proved exceptionally durable and is still used today as a warehouse and offices for the fair's plant engineering department.

Fair Park acquired a new tenant in 1909. The Dallas Art Association had outgrown a one-room gallery in the public library and offered its private collection to the city in exchange for a permanent home in the Fine Art and Textile Building.

Since the Coliseum could not be completed in time for the fair, the only additions to the physical plant were a Kennel Building for the dog fanciers and a replica of the Alamo built to one-half scale and presented as a gift from the *Dallas Morning News* and its general manager, George B. Dealey.

For all the problems of the preceding months, much excitement surrounded the 1909 exposition. The President of the United States, William Howard Taft, had decided to visit the fair on its middle weekend. The president was concluding a 12,000 mile goodwill tour, and detailed newspaper accounts followed his progress through the western states. Reports of an historic meeting with Mexican President Diaz were juxtaposed with speculation about Taft's failure to stay on his diet. With some distress, the already-corpulent chief executive sent word ahead to Dallas organizers requesting a menu of simple foods such as ham and eggs or corned beef and cabbage.

Miss Katherine Klarer, a splendid dramatic soprano soloist with Liberati's Band.

Taft was not the only celebrated visitor expected at the fair. Dan Patch, the greatest pacer of all time, was entered in an exhibition against a younger rival, Minor Heir. The champion traveled to Texas with a racing string that included two of his daughters. Another well-known guest, Comanche Chief Quanah Parker, brought two of his wives.

Educational exhibits included a collection of stoneware and pottery assembled by the residents of Athens to demonstrate the commercial possibilities of Texas clay. Fairgoers were also exposed to such unusual sights as a herd of Zebu, the hump-backed cattle of India; an Egyptian mummy; rare Pekinese dogs; and Prince Nicholi, a 22½″ midget who was hoping to barter his Russian title for marriage with a rich Texas woman. The prince's quest for a bride was covered extensively in the newspapers, and just when interest seemed to be dying down, someone attempted to steal the little fellow by stuffing him in a topcoat pocket. He was rescued, but a search for the kidnapper kept Prince Nicholi in the news for the remainder of his engagement.

The biggest attraction, apart from President Taft, was a series of flights over the park by a Strobel dirigible. The cigar-shaped airship was filled with hydrogen gas generated from its own engine. Aeronaut Frank W. Goodale sat astride the framework below the bag and steered a course by tugging on the rudder ropes. During one performance, the dirigible flew five miles to downtown and circled the 14-story Praetorian Life Insurance Building before returning to Fair Park.

Taft's train was scheduled to arrive on tracks near the main gate late in the afternoon of October 23. The area was roped off, and the president would be transferred to a motor car and driven to the grandstand for his speech. Members of the Texas National Guard were stationed along the route to provide

The first President of the United States to visit the fair, William Howard Taft — 1909.

Main plaza.

security. Shortly before the train pulled in, a tragic misunderstanding resulted in a guardsman fatally wounding a bystander with his bayonet. The victim was a respected deputy city clerk who simply had wanted to cross the line to catch a streetcar. Police arrested the young sergeant, and he was later convicted and sentenced to life imprisonment.

Taft was unaware of the incident. Wearing a silk hat and frock coat, the 300-pound president, who had a touch of laryngitis, raspily delivered a ten minute speech that could be heard by perhaps 5,000 of the 10,000 in the audience. He was honored at a banquet in the Oriental Hotel that evening, and the simple menu featured steak and vegetables. Dessert was plain vanilla ice cream — molded in the shape of possums.

On the closing day of the fair, Robert Burman drove his Buick to a new record of 101 minutes and 25 seconds for 100 miles around a circular dirt track. An elderly country doctor watched in disbelief. "The human race has gone mad," he said. "This will produce more strain than the anatomy can stand. In a few generations, men will be old at 30 and physical wrecks at 40."

Without racing revenue, the final receipts fell far short of the 1908 figure, but there was still a net profit of $31,000. The fair had survived and even prospered in the face of a loss that had nearly put it out of business only six years earlier.

1910

Once again, it was time to chart the country's progress with a census. Since 1900, the population of the United States had grown from 76 million to 92 million. Many of these were immigrants looking for the storied land of opportunity and processed through Ellis Island at a rate of 4,000 per day.

America had lost some of its beguiling innocence. Labor unrest, women's suffrage demands, prohibition rallies, socialist agitation and other movements tempered the bravado and easy confidence which had welcomed the 20th Century. But this was still a nation of ideals and heroes. Americans honored men of varied accomplishment: Ty Cobb, the base-stealing and batting star of the Detroit Tigers; George M. Cohan, a star of a different order whose hits were recorded on Broadway theater marquees; and Admiral Robert E. Peary, the Arctic explorer who documented his claim in a book *The North Pole* published in 1910.

There was, of course, a notable shortage of heroines, and the best-known black man in the country was Texas-born Jack Johnson, once a dishwasher in a Dallas cafe and now boxing's heavyweight champion. In 1910, the controversial Johnson knocked out another "Great White Hope," former titleholder Jim Jeffries.

Negro participation in the State Fair of Texas had declined over the past decade. Although they were still welcomed as visitors, Colored People's Day had been discontinued, and most blacks attended a separate week-long event in Fair Park which offered far more competitive classes than existed for them at the fair. Segregation was being legally defined in Dallas. The first "Jim Crow" law, which regulated seating in streetcars, was enacted in 1906, and one of the most deplorable incidents in the city's history occurred in 1910 when self-styled vigilantes broke through police lines guarding the courthouse, seized a Negro suspect and pushed him through an upper-story window into the hands of an angry crowd. The mob lynched him from a telephone pole beneath the ceremonial Elks' Arch over Main Street.

(right) The Coliseum, completed in 1910 at a cost of $108,000, was used for horse shows as well as musical entertainment.

(below) Downtown Dallas showing the ceremonial arch erected over Main Street for the Elks Convention in 1908.

The population of Dallas had climbed to 92,104, and one of the town's new businessmen was Robert Lee Thornton. Thornton had traveled a circuitous route to Dallas since his first visit to the State Fair as an Ellis County farm boy in 1889. His experiences as a day laborer, clerk, bookkeeper, candy salesman and now bookstore proprietor eventually launched a banking and political career which would have unparalleled influence on the future of Dallas and the State Fair of Texas.

The fair celebrated its Silver Jubilee Anniversary in 1910. Of the founders, only Gaston, Sanger and Sydney Smith remained active after 25 years, and only one of the permanent buildings pre-dated 1905.

The Coliseum opened with vaudeville acts from the Orpheum circuit and Thaviu's Russian Band in its first United States appearance. A box seating eight people sold for $50.

With the Coliseum available for musical entertainment, the wall that had separated the auditorium area from the Exposition Hall was removed. Uncle Sam was one of several new participants. The government built an aquarium for its fisheries exhibit and showcased the nation's naval strength with replicas of battleships, torpedo boats and submarines. A 50′-square natural history booth so impressed visitors with its educational values that teachers arranged to bring their classes to see it. The heavens were represented by a great cloth dome, and different varieties of birds hung from wires over a taxidermic jungle of wild animals.

Adolphus Busch sent a string of fine roadsters from St. Louis in addition to his brewery's famous six-mule team. The Goodnight Ranch sponsored a special barn for their Persian sheep, buffaloes and cattaloes, the latter a cross between a buffalo and a polled angus.

Dallas now had 40 legitimate automobile dealerships, and the latest models were displayed wherever room could be found in buildings and tents. A Maxwell Runabout was outfitted with flanged wheels and set up on a circular track under the grandstand to prove that this vehicle could run 24 hours, non-stop, for 16 days at a speed of 10 mph. Fairgoers were fascinated with this demonstration, but inhabitants of the nearby Indian Village suspected that evil spirits were responsible for the driverless car that kept them awake nights. The Maxwell dealer solved this problem by putting a 5′-tall white teddy bear behind the wheel, and the complaints stopped.

The main grandstand attractions were a massive military drill by 1,000 cadets from Texas A&M University and a three-day tournament that offered 40 knights on horseback competing with 10′ lances to impale rings suspended on posts around the course. Along the midway, youngsters clamored for parents to buy them "bouncers," paddles with rubber balls

(top) The new Dairy Building — 1910.

(below) A military drill by cadets from Texas A&M University.

By 1910, the city of Dallas was ready to acknowledge the need for formalized planning to regulate its growth pattern, which up to now had been determined largely by the wishes and whims of real estate speculators and developers. To keep pace with such progressive municipalities as St. Louis, Kansas City and Atlanta, city officials brought George Kessler back to Dallas to analyze existing problems and recommend a master plan for future development.

Kessler released his visionary design for Dallas the following year. The Kessler Plan cost $10,000 and proved to be the most effective consulting study ever commissioned by the city. The main elements incorporated a system of streets and parks, a central rail terminal, a belt railroad line, removal of existing downtown tracks, projects to straighten the Trinity River channel and build levees for flood control, a civic center and a broad boulevard linking Fair Park to downtown.

attached, or tiny whips. Another money-maker was a new game which featured a glib-tongued black man, clad in prison stripes, sitting above a tank of water and taunting fairgoers to throw at a target and try to dunk him.

Records fell in every area. Minor Heir set a new one-mile mark for a pacer on a Texas track. The largest one-day crowd, an estimated 125,000, poured through the turnstiles on the second Sunday. Total receipts topped $200,000, and educated guesses of the cumulative attendance were in excess of 900,000.

In 1911, the Park Board bought 7.4 acres on the eastern corner of Fair Park, completing expansion in that direction, and the State Fair of Texas built a Livestock Pavilion and a towering modern roller coaster. The $30,000, double-tracked coaster, located on the southeast side of the grounds, carried 16 cars over a mile-long circuit in 70 seconds.

Exhibitors contracted for all the usable space, inside or out. Even with added acreage and new facilities, the park was extremely crowded, a matter of concern for both the fair and the city regarding possible access for emergency vehicles. The disgraced Elks' Arch had been dismantled and moved to Fair Park, and the federal government converted its structural framework into a building for its naval exhibit. The University of Texas erected a tent to generate support for such state projects as institutions for the feeble-minded and ventilating systems in schools.

Fairgoers enjoyed deep sea diving shows performed in a mammoth tank, and they eagerly awaited a chance to sit behind the wheel of a 1911 Maxwell with fans blowing simulated breezes in their faces while scenery of Yellowstone Park was propelled past the windows. But the popularity

Aviator Cal Rodgers — 1911.

of land and sea exhibitions paled when compared to the excitement surrounding the first air show. Flights were scheduled to take off and return to the racetrack infield every afternoon. J.A.D. McCurdy flew his biplane as high as 500′ for as long as seven minutes until an abrupt landing in soft soil catapulted pilot and plane tail-over-nose. As a temporary substitute, management persuaded cross-country aviator Cal Rodgers to make a one-day stopover on his coast-to-coast flight, and finally Beckwith Havens was hired to finish the engagement. Havens was famed for altitude records and bombing demonstrations. To illustrate the aircraft's battle capabilities, Havens dropped oranges on selected ground targets while twisting the plane to avoid imaginary artillery fire.

A special day on the 1911 calendar saluted the Boys and Girls Corn Clubs, predecessors of later farm youth organizations, and another day honored SMU. Methodists from across the state had convened in Dallas to inspect the site selected for their new university.

Governor Woodrow Wilson of New Jersey, whose name was being mentioned as a possible presidential candidate, delivered an address to 6,000 partisans in the Coliseum on the final Saturday, and the fair closed a successful, though not extraordinary run. The biggest day drew 128,000, but overall attendance dropped below the previous year's level.

President Kiest announced his retirement, and the directors chose a former member of the Dallas Park Board, J.J. Eckford, to lead the organization in 1912.

New Jersey governor Woodrow Wilson was testing the waters of presidential candidacy when he spoke at the fair in 1911.

1912

James Joseph Eckford began practicing law soon after his arrival in Dallas in 1885. A respected attorney and former district court judge, Eckford was appointed to the city's first Park Board in 1905. He resigned three years later to take an active role in State Fair leadership.

Dallas continued to assume the shape of a major city. Among the construction projects completed in 1912, two would become landmarks: the Oak Cliff-Dallas Bridge, later known as the Houston Street Viaduct, and the last of Adolphus Busch's grand hotels, an ornate 18-story skyscraper appropriately named the Adolphus.

Fair Park improvements included a first-class, $35,000 restaurant row which replaced the wooden remnants of "Old Smoky," another rest cottage and a handsome fountain for the main plaza. The plaza was further enhanced by an ornamental

J.J. Eckford, State Fair President, 1912-1913.

A new restaurant row was built in 1912.

clock with an illuminated dial donated by Linz Brothers Jewelers. The newly organized Dallas Symphony Orchestra made arrangements to play a series of 26 concerts in the Coliseum beginning in November.

The make-ready process of each State Fair involved more than building buildings and announcing an opening date. Someone, most often the fair's secretary, had to book attractions, contract with entertainers, sell exhibit space and privilege rights, work out schedules, produce premium lists, organize special days and ceremonies, arrange for hospitality, place advertising, and supervise the efforts of more than 1,000 employees, participants and volunteers. In 1912, Sydney Smith died before plans for the exposition could be implemented. His assistant and son-in-law, William H. Stratton, assumed most of Smith's responsibilities.

One of the logistical problems facing Stratton centered on converting the track area in front of the grandstand into a football field for the three-day gridiron carnival scheduled immediately following the race meet. He also had charge of official welcomes for a diverse group of special guests including lecturer/philosopher Elbert Hubbard and American Federation of Labor vice president John Mitchell. The governor's luncheon was planned for the large cafe in the new restaurant row.

The first cooking demonstrations at the fair, which dealt with food preparation for children and old people, were presented in the University of Texas tent. Miss Jessie Rich, the home economist in charge of the display, used fireless cookers and iceless refrigerators in her program. The textile department showed more handwork than ever, indication to some that ". . . all women are not out reforming the world and fighting for their rights . . . womanly arts are still being practiced, exquisite examples of needlework, cooking, preserving and art should be encouraging to the man who still believes that outside interests only interfere with domestic duties."

William H. Stratton, secretary of the State Fair from 1912-1926.

The Parent Teacher Association sponsored an exhibit that featured model playgrounds, an ideal children's library of 500 selected volumes, plus lectures on child hygiene and Montessori teaching methods.

In addition to the three football games scheduled in front of the grandstand, Texas played Oklahoma on the field in Gaston Park next to the fairgrounds. The annual get-together of UT alums was held in the Coliseum and covered topics of interest to former students. One of the speakers, T.W. Gregory, chose the subject, "The University Gymnasium Is Important, But It Is Not All." As life would have it, a multi-purpose gym built on the Austin campus in the 1930s was named after Mr. Gregory.

The midway spotlighted Princess Victoria, a diminutive 22-year-old Australian native who stood 25½ inches tall and weighed 19¾ pounds. A few feet down the row, fairgoers discovered 20-year-old Miss Victoria, English-born, 5′1″ and 600 pounds, whose claim to fame was an arm that measured 31″ in circumference.

The 1912 State Fair of Texas hit the 900,000 mark in total attendance, set a new single day record of 140,000 and provided a comfortable bank balance to underwrite the next year's program.

1913

Fair Park's acreage was still not adequate for the numerous activities proposed for its year-round program. During the off-season, the racetrack infield was divided into four tennis courts, two baseball diamonds, storage facilities, parking areas and an experimental sorghum farm, but more space was needed. The logical direction for expansion was the 14-acre tract on the southwest, Gaston Park. W.H. Gaston and his children had developed this property as a private amusement park which included a small baseball stadium where the Dallas Giants played their schedule of Texas League games. The city wanted to purchase the park for $60,000, and the Gaston family agreed to accept payment in installments. Possession was another matter, since the Giants' owner, J.W. Gardner, had a lease running through 1916. The city offered to buy him out for $10,000, one-third of which to be paid by the fair, but Gardner held out for $12,000, and negotiations stalled.

A big civic parade with more than 1,000 municipal employees and animals from the city zoo opened the 1913 State Fair of Texas.

The fair's first Automobile Building.

Out-of-town visitors, who traveled to Dallas just once a year to see the fair, noticed that the State of Texas Game, Fish and Oyster Commission's hatcheries were now located on the grounds, and a large building had been erected for automobile displays. The auto show boasted 175 new models. Local dealers planned to give away a $500 car during the fair, and as further promotion, scheduled a society night in their building.

Auto polo was introduced in 1913. The sport, supposedly invented by two bored car salesmen standing on vehicle running boards and kicking a tin can, was played on a field which covered the width of the track in front of the grandstand. Teams, each consisting of a driver and a malletman, were matched in a 13-game series. The event was billed as "too fast for the movies."

The exposition calendar featured a Suffrage Day, the first recognized at the fair, and a Ragtime Night, which promised the latest popular hits as interpreted by Thaviu's Russian Band. The University of Texas conducted sewing classes and presented a medical exhibit which used an electrical apparatus to illustrate the flight of a fly from stable to table spreading germs along the way. The doctor in charge of the display offered a truism: "You can know a city by the alleys it keeps."

Society Night at the auto show was a formal occasion with dancing and entertainment. At 10 p.m., all the lights went off, the automobile head lamps turned on and the horns sounded while the band played "A Hot Time in the Old Town."

Boy Scouts came from 17 different Texas cities to celebrate their day, an occasion shared with the Camp Fire Girls who took special interest in the cooking demonstrations since that was considered an important step on the path to becoming a Mrs. Boy Scout.

Inclement weather during the second week held attendance to about 750,000, but the net profit was sufficient to fund another year's operation without concern by the directors.

1914

Judge Eckford stepped down from the fair's top post and was replaced by the general manager of the Associated Manufacturers of Cotton Seed Products, William I. Yopp. A Tennessee native, Yopp pioneered the cotton products brokerage business in North Texas.

In world news, the assassination of Serbian Archduke Francis Ferdinand touched off war in Europe in 1914, but President Wilson held firmly to a position of neutrality.

In events of local interest, Dallas was the smallest of 12 cities in the nation to be named a regional bank site for the newly-established Federal Reserve System. The bank served all of Texas and parts of Oklahoma, New Mexico, Arizona and Louisiana. A new city hall was completed, number five in the town's relatively brief history, and the Ford Motor Company opened a Dallas assembly line production plant.

Only one payment had been made on the most recent Fair Park acquisition, but the Gaston family released the city from further obligation, essentially making the property another in a long line of gifts to the community. Fair officials voted to make up the $2,000 difference between the city and baseball club owner, Gardner, a poor decision in retrospect since possession of the property was delayed until the end of the year and could have been achieved at no cost by waiting another 12 months for the lease to expire.

William I. Yopp, State Fair President, 1914-1915.

In an effort to rebuild attendance, the 1914 exposition was advertised as "A Different Fair," but it was difficult to attract attention with slogans when newspaper headlines on opening day proclaimed "Mexicans May Attack Vera Cruz" and "German Submarine Sinks Another English Cruiser."

A military tournament highlighted the list of attractions. The track infield was named Camp Bell and set up to accommodate 1,200 men and 400 horses. Daily programs featured drills, demonstrations of such combat skills as wall-scaling and bridge-building plus mock warfare exercises.

Lincoln Beachey, known as the "world's most daring aviator," was booked for a two-day show. His specialties were consecutive loop-the-loops, aerial tangos and cutting dollar signs in the sky with the trail of smoke from his small craft. Beachey strongly believed that the United States needed to form an aviation corps as part of its military program, and he included several novelty demonstrations to prove this point. One of these involved "blowing up" a 150′ model battleship by bombing it with melons and soda water bottles. His most spectacular stunt during the fair was climbing to a record altitude of 8,000′, then killing his engine and plunging straight down before pulling out at the last second.

A new baby event was inaugurated. The Better Babies Show, under the direction of UT's Miss Rich, was designed to have instructive and educational benefits rather than simply rewarding cutest dimples and engaging smiles. The event was scheduled over an entire week, and 500 babies were examined and given thorough physical check-ups by a team of doctors who also served as judges. Points were awarded on a scientific basis to select the most perfect baby.

The miniature railroad had been removed, and on-grounds transportation was delegated to modern, sightseeing buses which carried 17 passengers.

Entertainment was provided by Ewing's Zouave Band from Champaign, Illinois. Dressed in green jackets trimmed with yellow braid and gilt buttons, red trousers, white leggings, a

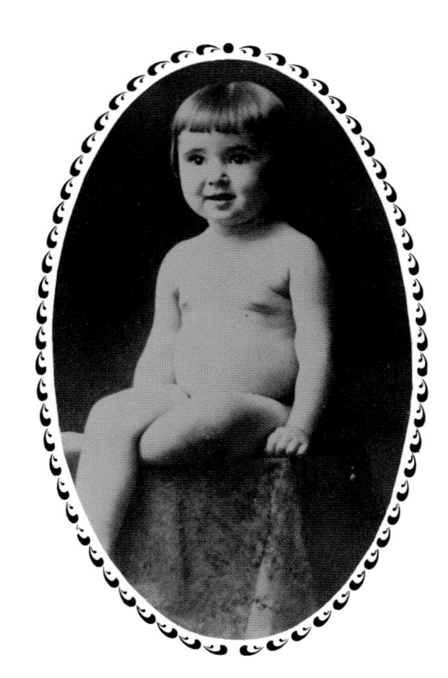

Judges declared two-year-old Grace Gulden the winner of 1914's Better Babies Show.

yellow sash and a red Turkish fez with a black tassel, this splendid group performed and drilled seven hours daily, but anticipated crowds never materialized. President Yopp described attendance as "considerably less than at previous fairs." Blame was placed on the war and its calamitous effect on the Texas cotton market, and Yopp requested that all courtesy pass holders pay their way through the gates on the final weekend. The appeal worked, and the exposition concluded with a razor-thin profit.

1915

Dallas continued to struggle with a depressed economy in 1915. The Park Board, looking for ways to finance upkeep for its various properties, went so far as to install coin-operated toilets in Fair Park's ladies' rest cottages.

Camp Bell was set up on the racetrack infield, and military tournament events were demonstrated for crowds in the grandstand.

With only one of its permanent buildings completed, Southern Methodist University began fall classes for about 500 students. "Birth of a Nation" was playing in local movie theaters when Texas governor James Ferguson arrived in town to open the 30th annual State Fair.

Camp Bell, temporary home of the 28th U.S. Infantry Division, again covered the infield area. Football games and a wild west show were scheduled on the track surface, and horse racing had finally been eliminated from the program. Although gambling had been illegal since 1909, and therefore officially nonexistent at the fair, facetious suggestions were made that betting men might want to redirect their wagers to how long aviator Art Smith might cheat the odds.

Smith replaced the late, and of course lamented, Lincoln Beachey, last year's aerial hero who failed to pull out of one of his spectacular dives at a later engagement. The "Ozone Wizard," as Smith billed himself, had enthralled audiences throughout the summer at the Panama-Pacific Exposition in San Francisco.

Heavyweight boxing champ, Jess Willard, who lifted the crown from Jack Johnson and thus received accolades as the "deliverer of the white race in pugilism," appeared at the fair in the 101 Ranch Show. Another star of the western extravaganza was William Eagle Shirt, a Sioux known as "the only Indian tailor," who overcame tribal scorn for taking up a squaw's occupation and created fanciful beaded and feathered outfits for his performances.

More conventional apparel, the latest fashions for the average woman as well as new French designs, was featured in a three-day style show in the Coliseum sponsored by *Vogue* magazine. The garments and 25 models came directly from New York for this first-time fair event. Highlights included a collection of auto coats, small plain hats and tea gowns raised a fashionable 10″ from the floor so that shoetops just reached the hemline.

Three pictures by Texas artist Julian Onderdonk were among 100 displayed in the Art Building, and the City of Dallas mobilized its offices for an impressive municipal exhibit. Charts and diagrams showed taxpayers where their money was going, and models illustrated new buildings and parks. In an adjacent booth, the state health department presented a chamber of horrors that rivaled anything seen on amusement row. A shelf held glass jars containing the horribly distorted brain of a meningitis victim, a section of intestines infected with hookworms, and various examples of dead mice and rats. A tombstone bore the engraved legend: "Sacred to the memory of 6,573 babies who died last year in Texas of preventable diseases."

The Agriculture Building showcased a new product, Texas ripe olives, and an exhibit entitled "Concrete on the Farm," which offered models of barns, silos, drinking troughs and milk houses made from that durable material.

For the first time in many years, a German-American Day was included on the calendar. A fine crowd turned out, and fair officials made special effort to commend the patriotism of Texas citizens with Germanic heritage. Celebrants talked about making it an annual event.

During activities that preceded the football game between Texas A&M and Haskell Indian School, the public got a first look at association football or soccer, a sport recently introduced to the Dallas area.

Halfway through the fair, most of the troops quartered on the grounds were dispatched to South Texas to suppress raids by Mexican bandits. The violent, ongoing struggle for governmental reform in that country had divided its revolutionary leaders, and the opposing forces of Venustiano Carranza and Francisco "Pancho" Villa were massed for battle near the Arizona bordertown of Douglas.

Art Smith was the sensation of the 1915 fair. He flew over the park at night, dropped fireworks from the plane, cajoled Mayor Henry Lindsley into taking a short flight and set a world record with 23 consecutive loops on the closing day.

Attendance bounced back to a respectable 776,260, and R.E.L. Knight was elected to lead the organization in 1916.

Silos and related farm equipment were displayed in outdoor exhibit areas.

In 1916, the Mothers' Council of Dallas started its new year righteously by urging the State Fair's directors to eliminate the sale of liquor on the grounds. The mothers expressed concern that young college boys were taking their first sips of alcoholic beverages during the annual exposition.

Somewhat cavalierly, Bob Knight and his board drafted a lengthy reply explaining that the fair was not a "camp meeting," that the loss of income would be critical and ". . . we believe it would be an admirable idea if the Mothers' Council would direct the affairs of that organization and permit the Directors of the Fair to run the Fair."

What began as a plea became a cause. Mothers circulated petitions. Ministers preached on "How to Run Booze Out of Fair Park." The Anti-Saloon League called a town meeting, and one of its leaders, Epps G. Knight, brother of the fair president, introduced a resolution condemning the sale of liquor in a public park and asking for a city referendum election on the issue. The outcome was never in doubt. Voters aligned themselves with motherhood, and a poem written by Henry Lamar for the *Dallas Dispatch* provided a rallying cry:

"The Fair will die!" The "fathers" have said,
The Fair will die without liquor red,
Such kind advice do the "fathers" give,
Telling the mothers that the Fair must live;
That the "fathers" run the great State Fair —
Mother's duty ends with household care!

In the palmy days of the gambling pack
When a halt was called in days of yore,
"The Fair will die!" It was said before,
That swarmed the grounds of the great race track
"The Fair will die!" Let the fever burn,
So long as the Fair big profits earn!

'Tis but an echo of former years.
But when this refrain comes to our ears
"The Fair will die!" Let the liquor run
And burn the throat of some mother's son,
And squander his coin and steal his pride —
Let no mother's son be here denied!
Tho it sear his brain and dim his eye,
Let liquor run, or "the Fair will die!"

"The Fair will die!" Well, then, die, say we!
'Tis sad 'twere so, but had better be;
'Twere better dead and right soon forgot
Than one mother's son a drunken sot,
In tears, we'll mark her honored grave:
"The Fair is dead — but she died to save!"

1916

Not surprisingly for the son of a Confederate officer, the R.E.L. in Knight's name stood for Robert Edward Lee. Knight became the first Texas-born fair president. His father had settled in Dallas County shortly after Texas statehood in 1845. A graduate of Southwestern University in Georgetown and the University of Texas, Knight was an outstanding lawyer and partner in a firm that served such clients as City National Bank, Sears Roebuck and Titche-Goettinger.

The passage of time was marked by the retirement of W.H. Gaston and the election of his son, Robert, to the fair's board of directors. The 75-year-old Gaston also stepped down as president of Dallas Electrical Light and Power Company in 1916, although he would remain active in business and banking for another ten years.

Attendance topped one million for the first time in 1916.

A portion of the fair's record first day crowd of 82,110 gathered for an afternoon ceremony to unveil the Sydney Smith memorial. The governor and other dignitaries participated in another dedication that morning which celebrated the opening of Union Station.

The rest of the fairgoers flocked to the sensational new amusement ride, the Whip, or hurried to see the light-armored motor car, a three-passenger, steel-plated vehicle with a revolving turret for a rapid-fire gun, already tested by the Army and Marines.

War consciousness was illustrated by a miniature city show called "Preparedness." This ingenious reproduction of a mythical coastal town underwent a fearsome mechanical and electrical attack, directed by two behind-the-scene operators, giving viewers graphic warning of what might happen to a complacent nation.

German-American Day disappeared from the program.

Other displays emphasized life's comforts and the improving economy. First-time exhibitors included Prairie View State Normal and representatives from the Rio Grande Valley. The latter brought carloads of fruit-bearing orange, lemon, grapefruit and banana trees and parrots which had been trained to talk about the virtues of the valley.

The local police department sponsored an exhibit featuring burglars' tools and weapons that had been used to commit recent Dallas crimes. Visitors interested in unusual livestock developments could check out the worm farm at the Texas Department of Agriculture booth or admire a pen of glossy black Karakules, the famed "fur" sheep from Bokhara Province north of India.

Art lovers compared paintings which had won gold medals at the recent Panama-Pacific Exposition with the works of

R.E.L. Knight, State Fair President, 1916-1918.

The Gulf Clouds Fountain, a massive bronze and granite sculpture, commemorates the contributions of the fair's first secretary, Captain Sydney Smith.

After Smith's death in 1912, one of Dallas' favorite young artists, Miss Clyde Gitner Chandler, was commissioned to produce a work in his honor. The $20,000 project was financed by private donations and through the sale of 50-cent "Sydney Smith buttons."

Chandler rejected the idea of a conventional statue. Her allegorical composition, intended to embody all the topographical and meteorlogical elements of Texas, was dominated by the enormous wings of a figure representing the gulf clouds. The artist wanted to pattern these wings after those of a seagull, but had difficulty obtaining a model since it was a federal offense to kill an ocean bird. She finally settled for a stuffed gull borrowed from the Chicago Field Museum.

The monument was erected on a plaza near the main gate. On October 14, opening day of the 1916 exposition, following a prayer by George W. Truett of First Baptist Church and speeches by the governor, mayor and State Fair president, Captain Smith's granddaughter, Isabelle Stratton, pulled a cord to release the American flag draped over the memorial.

For twenty years the sculpture greeted visitors as they entered the park. Then, prior to the Texas Centennial Exposition, it was loaded onto a truck and relocated near the 1925 auditorium which had been designated as a Centennial exhibit hall for General Motors. In 1972, the entire fountain was moved again, this time approximately 150′ across First Avenue to permit expansion of the auditorium, now known as Fair Park's Music Hall. This last move was a major engineering feat with a $40,000 price tag, twice the cost of Miss Chandler's original work.

such Texans as Reaugh, Onderdonk, Olin Travis and E.G. Eisenlohr, and commented favorably on the progress of southwestern artists. These critics ignored a painting displayed on amusement row. "Lady Godiva" was advertised as "a work of art in the nude" for the benefit of fairgoers not familiar with her ladyship's historic ride.

Auto racing returned, and a new athletic field located at the east end of the race enclosure hosted a full schedule of soccer and football. Players and coaches voiced appreciation for the field's soft soil, in contrast with the hard surface in front of the grandstand, and Texas-Oklahoma fans filled the 15,000 seats and crowded along the sidelines for a game that was gaining national stature.

The fair sponsored its first encampment for members of the Boys and Girls Corn and Canning Clubs from around the state. Youngsters were treated to a week of fairtime fun while

After a six-year absence, auto racing returned as a fairtime attraction in 1916.

attending short courses conducted by experts from Texas A&M.

Other than a few complaints about meat patty size in hamburgers, the popular sandwich introduced at the St. Louis World's Fair, the 1916 exposition ran a smooth course to its conclusion. Attendance on the final day pushed the total over the one million mark and solidified Texas' reputation as one of the most successfully operated state fairs in the country.

1917

"He kept us out of war," proclaimed supporters of President Wilson's bid for a second term, but by April 2, 1917, that was no longer possible. Able-bodied, red-blooded young men volunteered to fight "over there," and the rest of the country responded by purchasing Liberty Bonds, knitting woolens and collecting the peach pits necessary to manufacture filters for gas masks. It would be six months, however, before Pershing's men reached the trenches and fired their first shots.

In Dallas, construction began on a military airfield which would be named after a deceased flyer named Love, and preparations for the 1917 State Fair of Texas continued. A July fire destroyed the popular Royal Gorge ride, but arrangements were made to replace it with a similar tunnel ride called the Mountain Range. In late summer, the Park Board completed a large bandshell, and the fair stimulated interest in its second Boys and Girls Encampment by offering 750 scholarships to deserving rural youngsters.

Singer's Midgets, a 22-member Lilliputian troupe, headlined the Coliseum bill, and the Warthan Shows returned for another midway season. C.A. Warthan, known as "the little giant" by virtue of his diminutive stature and his clout as the biggest carnival operator in the country, brought an

The Hilton Sisters — Daisy and Violet — appeared at the 1917 State Fair. On June 18, 1936, during the Texas Centennial Exposition, Violet Hilton was married in the Cotton Bowl before 4,500 spectators. Daisy served as her sister's maid of honor.

unusual set of Siamese twins with him. Daisy and Violet Hilton were attractive little girls with considerable musical talent. Texas fairgoers would watch the twins grow up over the next 20 years.

Flags and banners in red, white and blue, hung from every available post, pillar and rafter. Bands played patriotic songs from early morning until ten at night, and the number one attraction was a huge British War Exhibit, a collection of mammoth guns, trench mortars, French periscopes, captured German naval mines, even uniforms and other gear belonging to the crack East Indian Lancer troops.

Fair visitors included Red Cross Rosie, in town to raise money for chocolate and tobacco to send to the front; U.S. Treasury Secretary William McAdoo, speaking in behalf of the Liberty Bond subscription drive; and Countess Kingston of Dublin, soliciting funds for disabled Irish soldiers. The Swine Breeders Association demonstrated their support by donating 27 registered hogs to be auctioned by the Red Cross.

Local patriotism was unfortunately linked with paranoia. Everyone suspected espionage when two children were hospitalized with intestinal disorders caused, in one doctor's opinion, by eating breakfast food which contained crushed glass. And persistent rumors forced architect Otto Lang, a city commissioner at the time, to publish a formal denial of spy charges.

The only other matter of concern to large segments of the population was prohibition. Dallas had voted dry on September 10, and the law was to go into effect during the fair. Looking for the eleventh hour windfall, several young men bought up large quantities of cheap liquor and attempted to resell it at Fair Park. The police interceded, and after October 21, drinkers were offered a variety of substitute beverages such as "Laperla," advertised as a satisfying nonalcoholic drink produced by the San Antonio Brewing Companies.

The fair finished with respectable numbers, and the directors discussed plans for 1918 including improvements in the livestock area, construction of a Press Building and proper annexation of the Gaston Park property which would involve setting the fence back and rerouting the streetcar tracks.

1918

Early in 1918, however, the war department expressed interest in converting Fair Park into a military training center. The Park Board and State Fair

Camp Dick — 1918.

Association offered the grounds. A portable YWCA building was erected near the Exposition Hall, and plans were made to use the Fine Art Building for dances and entertainment for soldiers. But in May, the government decided that it wouldn't need the property, and fair officials scrambled to organize a 13-day fall exposition. Shows were booked, agreements signed; then the government reversed itself again. This time fair management balked. Having already spent about $30,000 and now holding contracts, they insisted on some form of indemnification. At this point, the Dallas Chamber of Commerce, anxious to have the base and its perceived economic benefits, stepped in and agreed to reimburse the fair and cover its liabilities.

The army took control of Fair Park in late July and established an aviation boot camp. About 18,000 men were rotated through the center during its five months of operation, but less than one-third of these were officers, and the enlisted personnel had little money for discretionary spending with local businesses. The army planned to build a $10,000 swimming pool, which would have become one of the park's permanent assets, but the war ended before construction began. Camp Dick closed in December, and the property was returned the following April.

Those Roller Coaster Years

1919-1934

1919

The United States emerged from World War I as the most powerful nation in the world. Business boomed as consumers used wartime savings to buy homes, automobiles and other goods that had been in short supply. Veterans were assimilated into the job market, and industries expanded. Women finally won the right to vote. The jazz age had arrived, and the roaring twenties were just around the corner.

After repairing and cleaning up the mess Uncle Sam left behind, John N. Simpson, who was again president of the State Fair of Texas, announced a 14-day event for 1919. Exhibit space sold out completely two months in advance. The Dallas Automobile Trades Association built a 50′ x 400′ addition to the old frame hall, but even that proved inadequate to stage a show that now ranked with Chicago's and New York's as the country's largest.

Ride attractions at the "Victory Fair of 1919" included the sensational Whip and the always popular carousel.

An apple cider concession stand — 1919.

President Wilson's sudden, serious illness and the surprising World Series collapse of the favored Chicago White Sox were popular subjects for speculation when Governor W. P. Hobby opened the "Victory Fair of 1919" to the accompaniment of anti-aircraft guns and martial music.

The war was over, but certainly not forgotten. Carrier pigeons delivered messages, and a 35-ton tank climbed ditches, crushed trees and plowed through brick walls in demonstrations of frontline might and ingenuity. Grandstand patrons were treated to a re-enactment of the Battle of Chateau-Thierry. More than 200 ex-servicemen participated. Simulated artillery shells blasted the fortressed hill and the quaint French town at its base. According to one spectator, all that was missing was the blood on the battlefield.

Evening activities were enhanced by three giant search-lights, an opportunistic acquisition by fair secretary Will Stratton. The huge lights, originally intended for shipment to Russia, had been abandoned on a San Francisco dock after the Bolshevic revolution in 1917.

Traditional bands performed, but the top billed entertainers were the black musicians and vocalists of the American Syncopated Orchestra who introduced jazz to State Fair audiences. Special visitor groups in 1919 ranged from grizzled Confederate veterans to an infant organization formed for growers of sweet potatoes. Farmers from Louisiana, Arkansas, Oklahoma and Texas met for their first convention. Their banquet menu featured succulent yam delicacies for every course.

Ormer Locklear presented his daring Sky-Calisthenics on the final Saturday. As promised, the aviator scampered all over the wings and fuselage, hung by his heels from the landing gear and crawled down a rope ladder to change planes in mid-air. The record crowd of 161,790 was attributed partly to this spectacular aerial show. Most people came simply because the sun was shining. The postwar celebration had been soaked by ten days of rain.

1920

The fair finished in the black, but President Simpson noted that the organization had $70,000 to spend on a plant that needed $500,000 worth of improvements. Discussions centered upon new facilities for automobile and agricultural displays and a permanent football stadium, but the only projects completed were two livestock barns, fencing around Gaston Park and the installation of vandal-proof iron seats in the grandstands.

Edwin J. Kiest, State Fair President, 1908-1911 and 1920-1921.

In June, after a brief illness, Colonel Simpson died, and the directors chose another experienced leader, Ed Kiest, to replace him. Management adopted a slogan, "The Fair Without an Equal," returned to the familiar 16-day format and arranged for a prestigious Mexican art and industry exhibit. The now-stabilized government of Mexico sent its 110-piece Estado Mayor Band, and President-elect Alvaro Obregon agreed to make one of his first speaking appearances in the United States.

Two professors from the University of Mexico took charge of decorating the display areas under the tiers of seats in the Coliseum. Using canvas panels, bright paint and design patterns of Aztec origin, they transformed the drab, dark corridors into a brilliant showcase for 12,500 items of furniture, pottery, rugs, curios, cotton clothing and leather goods. The center gallery in the Art Building was devoted to 120 paintings by Mexican masters.

"Smiles of 1920" was the Coliseum stage attraction, and officials emphasized that this was a musical extravaganza along the lines of "Ziegfield's Follies" rather than the vaudeville entertainment of past years.

The fair had its first oil supply exhibit, and visitors rhapsodized over the model home built in Pan-American styling by Rodgers-Meyers Furniture. The house featured a glass-enclosed patio and side porches. Each room was

Texas counties advertised their agricultural accomplishments at the fair.

furnished to represent a different historic period. Prairie View Normal, which had grown to an enrollment of 2,100 students, presented an attractive display that explained its programs in teaching, farming, veterinary medicine, sewing, home nursing, auto repair and chauffeur duties.

Fairgoers, including Enrico Caruso who spent an afternoon in the amusement area, enjoyed Mack Sennett's Diving Girls from Southern California; the world's tallest man — a towering Dutch giant; football match-ups between SMU-Texas A&M and UT-Oklahoma A&M; and, perhaps most of all, sunshine. As in times past, the weather was exceptionally pleasant following a year of damp and cold, and the new electric turnstiles at the entry gates clicked a record 1,023,563 times.

1921

The initial burst of postwar prosperity was short-lived. Rising costs ate into the wage gains of industrial workers. Strikes and work stoppages triggered further inflation, more labor trouble and finally a brief, but severe depression. Prices fell, particularly for agricultural products, and unemployment soared.

The man charged with halting the economic slide and combatting widespread disenchantment was the 29th President of the United States, Warren G. Harding. The

An aerial photo of the postwar fairgrounds. On the lower right, the Gaston Park addition where the auditorium and swimming pool would be built in the mid-1920s.

(top) College football fans filled the 15,000 seats in the new wooden stadium built in 1921.

(bottom) A balloon ride is launched just outside the stadium entrance.

former Ohio newspaper publisher had pledged to return the country to "normalcy." Some interpreted this as a promise to roll back the clock and apply restraints to various radicals and rebels who were advocating communism or swilling bootleg gin. The reactionary climate spawned restrictive changes in immigration policy, a resurgence of religious fundamentalism, prohibition and a revival of the Ku Klux Klan.

There were indications of Klan activity throughout Texas in 1921, but on the whole, Dallas steered clear of the hard times and social malaise affecting much of the nation. The new Majestic Theater opened, radio station WRR began broadcasting and the latest and tallest skyscraper, the Magnolia Petroleum Building — at this point without its distinctive Flying Red Horse — neared completion.

Football fans followed construction of a 15,000-seat wooden stadium beyond the south end of the racetrack in Fair Park. To encourage greater participation in the boys' and girls' encampment programs, a dormitory was erected in the livestock area. Modern multi-purpose buildings replaced the flammable shacks on Parry Avenue opposite the main entrance.

Hoping to repeat the previous year's success, the directors again secured entertainment from Mexico and updated the Coliseum show to "Smiles of 1921." The football schedule called for games involving Baylor, SMU, Texas, Texas A&M and worthy opponents, but the opening night event in the new stadium was decidedly non-athletic. In what surely qualified as the earliest Texas Centennial pageant on record, Dallas celebrated not the Republic, not admission to the Union, simply the discovery and settlement of the Lone Star State. As intended by local retailers who sponsored the evening, society matrons bought elegant gowns to wear while seated in their reserved boxes, and the huge performing cast purchased materials and accessories for costumes. The stands were filled with ordinary citizens who came to watch the socialites watch the pageant.

The most interesting exhibit on the grounds belonged to Bell Telephone Company. Eight wooden horns were mounted on top of a tower in front of the main building creating a "remarkable voice-carrying apparatus," the first loudspeaker seen or heard in the southwest. This primitive public address system had been introduced at Harding's inauguration and was used at the fair to page lost children, broadcast music and announce World Series scores.

A miniature biplane ride — 1921.

Noise amplification was neither necessary nor desirable along the midway. Tom-toms, triangles, deep sea fog horns, cow bells, drums and squawkers vied with sideshow spielers for fairgoers' attention. Carnival boss, C. A. Warthan, constructed one of the world's largest merry-go-rounds as a permanent park addition and installed an aerial swing fitted with miniature biplanes.

Though improvements in motorized transportation fascinated everyone, horse shows also remained popular. Visitors showed renewed interest in saddle horses for afternoon gallops around town or in the suburbs.

Kids thronged through the gates on Kids' Day to take advantage of free admission, free shows and free ice water. A record 2,800 youngsters reported themselves lost and had their names announced on the loudspeaker.

Motorcycle races, Alaskan ice skaters, the Singing Hussars, and an All-American art exhibit from Taos, New Mexico, were presented under sunny skies, but crowds were comparatively sparse in the fall of 1921. Attendance dropped precipitously to 650,000.

1922

"Inadequate and unsafe" were the words new president Harry Olmsted used to describe the aging Automobile Building. To construct its replacement, the fair arranged to borrow funds from the Dallas Automobile Association and the Dallas Manufacturers' Association. The structure would cost $150,000 and serve the combined needs of the two organizations.

Candidates backed by the Ku Klux Klan carried Dallas County elections that spring. Several prominent ministers supported the movement, and among city officials, only Mayor Sawnie Aldredge voiced public opposition. Dallas Klan 66 held meetings in both the Coliseum and Livestock Arena during the off-season, but the Park Board, reacting to the

Harry Olmsted, State Fair President, 1922-1924.

increasingly volatile social climate, refused requests by Negro teams to use Fair Park baseball diamonds, even though this had been accepted practice for years.

Radio was the rage of 1922. About 60,000 families in the United States already owned sets, and the fair organized a special showing of improved receiving equipment. Dallas station WDAO obtained permission to relocate its operation, and using two 80′ steel towers, originated live daily broadcasts from Fair Park through the run, a period reduced to 10 days by management in a budget-tightening move.

The new Automobile and Manufacturers' Building, measuring 185′ x 500′, was finished just in time for the opening. Floor space was apportioned on a 60:40 basis in accord with the funding raised by each trade association.

Rescue demonstrations by a crack crew from the United States Bureau of Mines highlighted grandstand entertainment. The team came to Dallas directly from the Argonaut mining disaster in California. Evening programs featured "Singsongs," with the words of old favorites flashed on two big screens to encourage audience participation, and daytime fare included auto racing on a track well-oiled with Texas crude. In one of the early races, driver Ben Gotoff suffered serious injuries when he failed to pull out of a skid, hit the wall and catapulted into the air with the car landing on top of him.

Further excitement was provided by a 19-year-old Tennessee girl, Lillian Boyer. From a speeding car, the young aviatrix caught a rope ladder attached to a swooping plane, climbed up to the wing and hung on while the plane soared and looped.

Exposition Building visitors gravitated to the Volks booth to try a new X-ray machine designed to check shoe fit. Remington exhibited a giant typewriter capable of printing on paper more than 2½′ wide, and Linz Jewelers showcased a ragged Count of Monte Cristo on his knees in a cave that sparkled

Stunt flyer Lillian Boyer was the sensation of the 1922 exposition.

The Mistletoe Creamery exhibit showcased a cow and milkmaid sculpted entirely out of butter.

with diamonds, rubies, sapphires, pearls and emeralds. Four guards stood by to protect the gems, valued at $250,000.

The newest midway rides were the Butterfly, which lofted its passengers along an erratic aerial path, and the Broadway Whirl, a spinning platform that challenged the rider's grip and equilibrium. Barbequed chevron captured the spotlight as the featured edible on Restaurant Row. Chevron was a polite term for goat meat. The name had been selected out of 2,500 entries in a contest sponsored by the Texas Sheep and Goat Raisers' Association, which was looking for a socially acceptable word along the lines of beef or mutton. The winner's prize, as might have been expected, was a goat.

Nearly 700,000 attended the shorter exposition, an encouraging sign which led the directors to plan a 16-day event for 1923.

1923

Louis Blaylock headed a ticket endorsed by the Ku Klux Klan that swept the municipal elections the following spring. Dallas' new leader, a wealthy Baptist publisher, had served competently in various elective positions prior to assuming mayoral duties. Klan influence in city government now compared to that of a political machine, and Mayor Blaylock was an open friend, if not an actual member of the organization.

In 1923, local residents were saving money by shopping for groceries at Piggly-Wiggly, a rapidly-expanding national chain that had pioneered the self-service system. By cutting personnel costs and purchasing in volume, Piggly-Wiggly was able to underprice even its larger competitor, A&P.

Two marketing innovations of the 1920s, chain stores and installment buying, played significant roles in the economic spiral that was beginning to take hold across the country. Equally important, President Harding and his successor, Calvin Coolidge, personified the laissez-faire approach to government regulation of business.

Harding had proven a weak and indecisive chief executive. Rumors of scandal and blatant corruption within his administration were just surfacing when the president became ill while on a western speaking tour. He died in San Francisco on August 2, thereby elevating the taciturn Coolidge to the nation's highest office.

In October, Dallas celebrated the reopening of Pacific Avenue as a downtown traffic artery without railroad tracks and the return of horse racing — still without betting — to the State Fair. New stables had been built, but union leaders were unhappy that Mexican labor had been used both for this and construction of the Automobile and Manufacturers' Building a year earlier. They asked their members to stay home from the exposition and threatened to call a unified protest by all labor unions, including the musicians and stagehands, in 1924.

Fair visitors rated racing, rodeo and Art Landry's Syncopated Chicago Jazz Band as their favorite attractions. They also took notice of "Little Tom," a 3,000 pound ox pulling the Budweiser wagon, and the first scientifically authenticated mother mule, on display at the Texas A&M tent. These natural wonders competed with the latest model Columbia Graphanola, offering an automatic start-and-stop system and improved sound reproduction, and balloon or

Local dealers displayed their newest models at the State Fair Automobile Show.

"doughnut" tires seen for the first time at the automobile show. Manufacturing exhibits featured action displays of hosiery knitting and paint grinding, and metal beds with a wood-like appearance were introduced to Texas by the Simmons Company of Kenosha, Wisconsin. One enterprising automobile dealer allowed a trio of black musicians from nearby "Deep Ellum" to perform in his exhibit area. An appreciative crowd gathered, and the next day other dealers were out on Elm Street trying to book entertainment.

Guest speakers at the 1923 exposition included ex-Governor Frank Lowden of Illinois, Senator Royal Copeland of New York, Senator Oscar Underwood of Alabama and Hiram Evans, a former Dallas dentist who had risen to the highest post in Ku Klux Klandom.

Ku Klux Klan Day — 1923.

The State Fair of Texas designated October 24, 1923, as Ku Klux Klan Day. The official agenda began that morning in North Dallas at the dedication of Hope Cottage, a two-story home for "unbidden babies," built and donated by Dallas Klan 66.

The Dallas Klan claimed 10,000 members and was considered one of the most powerful in the United States. Arrangements had been made to bring Klansmen to the fair on special trains from all parts of Texas and Oklahoma, and some predicted a single day record attendance approaching 300,000.

Upon his arrival in Dallas, Imperial Wizard Hiram W. Evans emphatically denied Klan involvement in floggings or tar-and-featherings. At the same time, he glossed over past differences and praised the *Dallas Morning News,* although the newspaper had run a long series of articles exposing Klan practices. But when he stepped to the loudspeaker that afternoon and looked out on an audience of thousands massed in Fair Park's main plaza, Evans preached basic Klan doctrine. He spoke on: "Immigration Is America's Big Problem" and noted that the success of the United States rested on its ability "to fuse into a true Americanism the various nationalities that have flown into the country through immigration, casting aside as dross those materials that cannot be assimilated." Then he got to the point and cited three groups that were "absolutely unblendable" — Negroes, Jews and Catholics.

Throughout the day, fairgoers sported KKK badges, and some of the rodeo cowboys, including all-round champion, Yakima Canutt, wore Klan regalia and robes while performing. At 9 p.m. in front of the grandstand, masked Klansmen initiated 5,631 candidates into the organization before a supportive crowd of 25,000. Crosses burned as crusaders sang, "Onward Christian Soldiers," and the festivities ended with a downtown parade. The final turnstile count was around 150,000, somewhat less than expected, but the occasion generated tremendous publicity for the Klan and considerable revenue for the fair.

Ku Klux Klan Day did not become an annual fairtime event. Before the next exposition, leadership squabbles had splintered the local Klan, and similar power struggles in other communities and at the national level eroded the organization's effectiveness as a political force.

1924

For several years the City of Dallas had recognized its need for a large auditorium, and after giving consideration to a downtown site, a decision was made in 1924 to build the hall inside the fairgrounds. City ownership of the land and adequate provision for parking dictated the selection of Fair Park. Architects Lang and Witchell submitted an imaginative proposal for a facility contoured in the shape of a megaphone, a radical departure from the conventional box-like design. The structure would cost $500,000 and seat 5,000 patrons.

With an auditorium in its future, the State Fair filled an immediate need by converting the Coliseum from a substandard music hall to an altogether satisfactory Agriculture Building.

Using the same means that had provided funds for earlier construction, the fair asked 20 major exhibitors to put up $1,000 each to finance extensive remodeling of the Exposition Building. Another project scheduled for completion prior to the fair was a new 2,000′ roller coaster called Lightnin'.

Preparations for the exposition were not without problems, however. The "closed shop vs. open shop" dispute between the State Fair and local unions erupted again, and the Dallas Central Labor Council declared a boycott on the fair. Then an epidemic of hoof-and-mouth disease broke out in South Texas causing federal and state authorities to cancel the annual livestock show. A later reprieve permitted exhibits of horses, mules and poultry only.

The 1924 fair featured the traditional fireworks spectacle, as always, themed around a horrible disaster of worldwide scale. This one was titled "Tokyo" and modeled after the earthquake and fire which had nearly destroyed the Japanese capital in 1923.

As the high point of Magicians' Day, Harry Houdini addressed a large grandstand audience on the topic: "Can the Dead Speak to the Living?" Houdini exposed various charlatans who promoted the cult of spiritualism. After the lecture, he entertained the crowd by wriggling out of a strait jacket while hanging upside down. Houdini's efforts to debunk the spiritualists' claims did not stop him from setting up an experiment to be carried out after his death to explore the possibility of communication beyond the grave.

The $20,000 renovation of the Exposition Building won praise from exhibitors and fairgoers alike. A Spanish Village had been constructed within the cavernous hall. Quaint little shops, some with second level porches or balconies, replaced booths; aisles became streets; and a fountain which duplicated a familiar tourist sight in Barcelona stood on the central plaza. Ed Kiest donated a set of cathedral chimes to hang in the village tower.

(top) Lightnin', a spectacular new 3,000′ roller coaster, was built in time for the summer season in 1924.

(below) The interior of the Exposition Building was renovated with a Spanish Village theme.

Lightnin' started rolling at sun-up each morning. The first drop plunged 60′, and riders agreed the thrills were worth the 25-cent charge. The faint-hearted could also enjoy the coaster craze on Lightnin' Junior or the still formidable Mammoth Racer. Those with queasy stomachs stuck to the Old Mill, a leisurely boat trip through the canals of a picturesque Dutch village.

Railway and newspaper unions declined to participate in the Labor Council's boycott, but members of the carpenters' and stagehands' unions were fined for each fair visit, and Council pressure finally forced Al Sweet and the Singing Hussars to break off their engagement or face union expulsion.

Evening entertainment after the SMU-Texas football clash called for an All-Collegiate Circus with schools from every section of the country participating in a competition to determine the best band, variety act, tumbling team and campus beauty. An Oklahoma coed was crowned queen.

During the last week, World War I ace, Lt. Henry Toncray, piloted his "flying automobile" — an invention that crossed a car with an autogyro — and the U.S. Navy dirigible "Shenandoah" flew over the park. Attendance once again fell just short of one million, but all things considered, the directors rated 1924 as a very good year. With Coolidge in the White House assuring everyone that, "the chief business of the American people is business," fair officials were anxious to begin construction of the long-awaited auditorium.

1925

Work on the theater got underway in January of 1925 and proceeded with a minimum of controversy. The Labor Council viewed this project as separate, and therefore exempt from its continuing boycott of the fair. Questions arose about fire-proofing protection that covered only a portion of the building. The architects provided an interesting statistical explanation showing that most auditorium fires started in the stage area, and by the time the blaze reached the seats, the audience would either be out of the building or dead from smoke inhalation.

The Park Board settled their debate about a name for the theater by throwing the matter open to the public. A cash award was offered for the best suggestion which turned out to be the most obvious: Fair Park Auditorium. Alex Sanger proposed the motto, borrowed from an inscription over the entrance to the Frankfurt Opera House: "Dem Wahren, Schoene, Guten," or "To the Pure, the Beautiful and the Good."

A.A. Jackson, State Fair President, 1925-1926.

Newspaper publisher Edwin Kiest pledged a generous $5,000 toward purchase of a $50,000 Barton pipe organ for the hall. The emerging structure, with its polyglot of arches, turrets and domes, reminded some observers of an enormous Spanish mission. Others, perhaps charitably, avoided any comment regarding its style.

In June, the fair's new president, A. A. Jackson, announced that a contract had been signed with the Shubert organization of New York, the most respected theatrical producers on the Great White Way, to bring the Broadway hit "Sky High" to Dallas in October as the auditorium's first show. Comic Willie Howard headed a 108-member cast for this lavish musical production which had filled New York's Winter Garden Theater for months. Overlooked in the excitement surrounding "Sky High" was a brief but ominous report that Al Sweet and the Singing Hussars had paid a $3,000 fine to be reinstated in the musicians' union.

While the musical would be the exposition's headliner, other entertainers and exhibits were booked to round out the program for 1925. Workmen laid a special spur track to accommodate the Blacklander, an agricultural display train from the Gulf Coast, and a modern funhouse was erected on the midway. This attraction featured all the devices and torments popularized at California beach resorts — an obstacle course past bumpers and barrels, whooshing air jets, slides, trick mirrors and the frustrating fox trot.

Federal quarantine authorities continued to bar sheep, swine and cattle from the livestock show, so the fair directed attention to the equestrian events by inviting a well-known public figure and expert horseman, Colonel Billy Mitchell, to umpire polo matches and judge the hunters and jumpers. Mitchell, a flying hero turned war department critic, agreed to participate pending the outcome of court martial proceedings against him in Washington.

The magnificent pipe organ was delivered in September, and the fair scheduled a special recital by virtuoso Clarence Eddy and daily concerts by Ralph Waldo Emerson, staff organist for Radio Station WLS in Chicago and a lineal descendant of the well-known essayist.

Fairgoers — 1925.

Two weeks before the opening, union representatives tired of stalled negotiations with the fair and threatened to strike every Shubert theater across the United States if "Sky High" played in Fair Park. Fair directors and city officials met privately. Rumors circulated that the Shuberts had asked to be released from their contract and wrestling matches might be offered as substitute entertainment. An intermediary initiated talks with the Dallas Central Labor Council, and on September 30, J. J. Shubert told reporters he was privileged to announce a new show for Dallas, "The Student Prince." Fair officials expressed their delight and explained that this was the

Fair Park Auditorium, built in 1925 at a cost of $500,000.

show they had wanted in the first place. Next came word that agreement had been reached settling the three-year quarrel between the Labor Council and the State Fair. Union leaders said they looked forward to working with fair management. All the closed door dealings and this sudden outburst of happiness and harmony left the public so confused that no one rushed out to buy tickets for "The Student Prince." But glowing reviews stirred activity at the box office, and Sigmund Romberg's melodic saga of Old Heidelberg became a standard against which later shows would be judged.

The 1925 fair also boasted the first Texas Industrial Exposition presented by manufacturing interests from around the state and the first fairtime football game between Negro schools. Punters from Wiley College of Marshall and Oklahoma's Langston College had a busy day as the teams battled to a 0-0 tie.

Attendance totaled 731,933, well below the 1,023,563 mark established in 1920, but the net profit exceeded $100,000, making this the most successful fair in financial terms since 1908. Under an arrangement with the city, the fair had promised to repay $300,000, plus interest, of the $500,000 spent constructing Fair Park Auditorium. The first $30,000 installment was delivered cheerfully and on time.

1926

The Park Board was prepared to spend more money in Fair Park in 1926. All parties agreed on the next project, a swimming pool, but opinion was divided as to where it should be built. Fair officials pressed for

Despite good crowds on selected days, overall attendance dropped to 542,872 in 1926.

Court proceedings grabbed most of the headlines in October of 1926. While exhibitors and officials at the State Fair of Texas waded through a week of rain, stay-at-homes found plenty of vicarious excitement in the newspapers. Charges of assault leading to the death of a young girl had been lodged against an Indiana Klan leader, and testimony continued in the sensational kidnapping case involving the charismatic California evangelist, Aimee Semple McPherson. On the local scene, courthouse gossip revolved around the upcoming murder trial of well-known Fort Worth minister J. Frank Norris.

construction south of the racetrack near Amusement Row. After listening to these arguments, the Park Board located the $75,000 recreational facility along Grand Avenue in the general vicinity of the Automobile Building and new auditorium. Built above ground, in a design that incorporated dressing rooms into the sides of the structure and used the concourse as a roof for the locker area, the circular pool accommodated 600 swimmers.

The city also decided to extend Grand Avenue through the park to the eastern boundary. By doing so, the street would slice across the northern end of the racetrack. The fair raised no objections, since the directors had already discussed reducing the track to a half-mile oval. Though a popular attraction with fairgoers, horse racing continued to operate at a financial loss, and the fair ultimately chose to abandon the entire program in 1926. The board voted to set up carnival shows on the infield and use the area segmented by Grand Avenue for parking.

Earlier in the year, a freakish hail storm caused extensive damage to several park structures, in particular the glass-domed Fine Art Building, but by opening day repairs and refurbishing were completed. Attendance had averaged around 800,000 for the past ten years, and President A. A. Jackson, a respected produce wholesaler with more experience counting cabbages than people, stated confidently that the 40th annual exposition would draw one million visitors.

To enable families to wander through the buildings and peruse merchandise from armadillo baskets to air-cooling systems, the fair set up a check booth for babies. Parents brought an infant to the entrance; a numbered check was issued and hung around its neck and mother kept the stub until it was time to reclaim the child.

"Princess Flavia," a musical version of "The Prisoner of Zenda," entertained theater audiences. Afternoon programs featured Thaviu's Russian Band and a massed choir formed by members of Dallas vocal clubs. Professor Thaviu, who had been the premier attraction at the first Coliseum show in 1910, admitted to being "mystified, puzzled and discouraged" by all the jazz around him. Bowing to the inevitability of change, he accepted his reduced station and stoically accompanied the amateur chorus.

More in tune with the times was an act called "The Charleston Steppers." Using living art poses and electronic effects, this all-girl show reproduced the paintings of Old Masters in tableau form. For a grand finale, the figures came to life and danced a frenzied Charleston. Another crowd-pleasing midway performer was Napoleon, the talented chimpanzee used by Clarence Darrow to demonstrate the plausibility of evolution at the Scopes Trial in 1925.

Fountain service at the Dr Pepper booth enabled fairgoers to enjoy the popular soft drink at 10-2-4.

Entries in the fair's cake competition reached an all-time high. According to the superintendent, Miss Fannie Howard, "It just goes to show that a modern girl is not forsaking the home arts, even if she does play golf and tennis . . ."

A Parkland Hospital medical supervisor, Dr. H. C. Standifer, commented that some of his patients were making remarkable improvement in order to go to the fair. He added that the hospital always had a surplus of empty beds at fairtime.

The cake contestants and recovered patients didn't have to fight the usual crowds at the 1926 exposition. Only 542,872 of the faithful showed up. Secretary Will Stratton grumbled about the rain, the price of cotton and the costly storm damage when he filed his report showing net profits down $50,000 from the previous year. The fair was unable to pay its second installment on the auditorium debt.

Rumors of mismanagement surfaced. At the annual meeting, a group of stockholders proposed a rival slate of nominees against six directors standing for re-election. This had never happened before, and the candidates supported by the disenchanted faction won. When the newly-constituted board met, Louis Lipsitz, managing partner of Harris-Lipsitz Lumber, was voted into the presidency. Various vice presidents were elected, but the office of secretary was left vacant. Ten days later, W. H. Stratton, a 13-year veteran as the fair's top salaried manager, submitted his letter of resignation which was accepted "with best wishes for Mr. Stratton, wherever he may go." The board overruled the executive committee's recommendation and granted Stratton four months severance pay. Chief clerk Roy Rupard was appointed acting secretary.

Four more directors, including outgoing President A. A. Jackson, resigned in the first few months of the new year, and the publicity manager also quit. Lipsitz obtained permission to have the last two fairs' financial records re-examined, after which the matter apparently was closed.

Stratton accepted a position with H&B Beer Company. His death, a result of injuries sustained when he was struck by a car, received front page notice in 1933. Most of the pallbearers at his funeral were former State Fair directors.

1927

Sadness and misfortune dogged the organization in the early months of 1927.

Louis Lipsitz, State Fair President, 1927.

The fair's patriarch, William Henry Gaston, died on January 24 at the age of 86. An editorial in the *Daily Times Herald* noted: "Few men have served their city as nobly and as effectively as Captain Gaston."

In April, new president Louis Lipsitz collapsed and died while attending a roof garden dance at the famous Crazy Hotel in Mineral Wells. Backed by a unified board of directors, Harry Olmsted agreed to serve out Lipsitz's term.

From the beginning, much of the exposition's success had been built on a broad base of competitive activities. In an effort to get the fair back on course, the directors announced new contests, new prizes and new spectator events.

Though the best-known sportswriter of the day, Grantland Rice, had coined a memorable couplet, a verse which insisted that the Great Scorer was less concerned with who won or lost than with how the game was played, most Americans were enamored with first place finishes. The stockmarket had turned small investors into winners — at least on paper. Charles Lindbergh conquered the Atlantic in 1927. Babe Ruth swatted 60 home runs, and movies "talked" when Jolson sang in "The Jazz Singer."

The State Fair latched onto the free-spending, euphoric mood by introducing whippet racing and bringing back the ponies for an expanded program on a rebuilt ¾ mile track. The directors promoted a hog-calling contest, a barnyard golf tournament, even a competition for teams skilled in the assembly of airplanes. The Chamber of Commerce cooperated by offering cash awards for the best costumes worn on Dallas Day and encouraging former residents to come back for an Old Home Celebration during the fair.

A crowd of 20,000 gathered on the plaza in front of the Exposition Building to watch Dallas mayor R.E. Burt cut a cake baked in honor of the city's 71st birthday.

Catching the spirit, Mayor R. E. Burt decided that the city would mark its 71st year with a party and brobdingnagian birthday cake in front of the Exposition Building on Dallas Day. And the Marsalis Park Zoo loaded a sampling of Texas-born animals, including lion cub triplets and an albino polecat, into horse-drawn wagons and trekked the menagerie across town to give the exposition an authentic wildlife exhibit.

In an odd note, Texas fairgoers had an opportunity to select a monument for their neighbors in Oklahoma. A wealthy benefactor wanted to build a 70′ statue of a Pioneer Mother on a knoll near the national highway running through the Cherokee Strip. Twelve of the nation's finest sculptors submitted designs. At a New York showing, the public cast votes for a favorite. The model chosen by New Yorkers provoked howls of protest from westerners, so a second showing and vote were scheduled at the State Fair.

Culturally-attuned visitors loved the Auditorium feature, "The Countess Maritza." Others came to the exposition to see the world's largest Victrola, daytime fireworks and a prototype of death row's electric chair.

Aspiring musicians were invited to audition for the chance to make a phonograph record. Tryouts were held each afternoon. Winners furnished music for the daily public dance in the Exposition Building, an event which was broadcast over Radio Station KRLD from 5-6 p.m.

Henry Toncray, the stunt-flying lieutenant at the 1924 exposition, returned to demonstrate crop dusting methods for cotton farmers. Toncray skimmed over the infield leaving a white cloud in his wake. Fortunately for those below, the plane spewed a harmless white powder rather than the calcium arsenate intended for boll weevils.

The weather remained warm and sunny through most of the two week period. Attendance climbed back to the one million level, while fair officials grappled with the usual kinds of problems. Texas football fans registered complaints about their ticket allotment for the big game against Vanderbilt. The

Participants in the Boys and Girls Encampment at the State Fair in 1927.

city health director dismissed 40 food handlers for lack of personal cleanliness. The city attorney charged that gambling was probably taking place at midway game booths, but admitted he had no proof. A circus elephant dropped dead and left a large 18-month-old orphan. And a potentially serious fire in the livestock area was averted by a quick-thinking maid who extinguished the blaze with a tub of soap suds.

1928

A murmur of criticism about the acoustics in Fair Park Auditorium prompted some corrective remodeling in 1928. Other patrons groused that the hall's beverage service was noisy and slow. One man suggested that management could pass out free ice water at intermission, a courtesy extended by Karl Hoblitzelle's theater in St. Louis.

But no one was complaining about the first-rate attractions being offered to local audiences. Fair Park Auditorium had put Texas on the map in the minds of New York producers.

Alexander Gray starred in "The Desert Song."

Fair officials were learning the intricacies of theatrical negotiations. No longer willing to abide by the decisions of paternalistic New Yorkers, the directors took a firm stand in the spring of 1928. The show they wanted was Sigmund Romberg's latest operetta, "The Desert Song," and President Olmsted threatened to keep the theater dark in October rather than accept anything else.

The strategy worked. "The Desert Song" was a $90,000 production, the most expensive ever brought to Texas. As word filtered back about block-long lines for the Los Angeles engagement, area residents hastened to buy tickets. The top price had risen to $4. First night seats were a "must" in the best social circles.

Movie comic Ben Turpin came to town to kick off the sale of "Opening Day Keys." This advance purchase promotion, sponsored for several years by the Junior Chamber of Commerce, offered tiny keys or buttons that entitled the bearer to admission and special discounts.

Prior to opening, the fair converted restaurant row, where church organizations had operated food booths since 1912, into a home for the growing poultry department. The old Poultry Building became the new Dairy Building. Over the years, this practice of continually remodeling and renaming its facilities permitted the fair to change with the times and capitalize on opportunities. Often, however, this made the word "new" an ambiguous term in descriptions and records. Map-sellers reaped annual benefits when visitors returned each October and looked in vain for familiar landmarks.

Farmers were invited to confer with superintendents and county agents during their fair visits in 1928. The reorganized Southwest Dairy Association presented its first show, and hitching demonstrations illustrated the effectiveness of using three, four, six, eight or ten horse teams rather than the traditional two.

The famous Curtiss Carrier Pigeon, the liberty-motored biplane that inaugurated air mail service between New York and Chicago in 1926, hung from the ceiling of the Automobile Building, and a huge auto maintenance exhibit and clinic occupied one corner of the hall. Lack of space limited dealers to displays that did not include test drives.

Aviation exhibits were displayed in the Automobile and Manufacturers' Building in 1928.

German Day appeared on the calendar for the first time since 1915. Czech Day had been established a few years earlier and was enthusiastically supported by small Texas communities with middle European roots.

Management booked Cecil B. De Mille's epic film, "King of Kings," for the Auditorium on afternoons when there was no operetta matinee. Another successful attraction was the Photomonon, a million dollar camera that could take a strip of individual pictures in 15 seconds and produce the finished print in just eight minutes. Even more popular was the live lizard concession. Young men, prone to goldfish swallowing and other fads of the '20s, jauntily wandered the grounds with chameleons clinging to their clothing — foreshadowing, perhaps, the day when status would be defined by tiny reptiles embossed on shirt fronts.

Harry Olmsted concluded his fifth term as State Fair president and handed the organizational reigns over to the Southwest District Manager of Pittsburgh Plate Glass Company, T. E. Jackson. After two consecutive years with attendance above the rarified million mark, board members were ready to give serious thought to a new football stadium.

1929

The nation's 31st president, Herbert Hoover, rode a tidal wave of prosperity into office. Prices on the New York Stock Exchange, having reached historic levels in 1928, climbed even higher in 1929. Record earnings, low taxes and a mania for speculation fed the market surge.

Reflecting the mood of the country, State Fair leaders started planning an 80,000-seat stadium expected to cost at least $750,000, even though they had not finished paying their share of the auditorium debt. Indeed, despite two outstanding financial years, the fair had yet to catch up on the payment it

T.E. Jackson, State Fair President, 1929-1931.

had missed, and the Park Board finally authorized the city attorney to write a letter threatening to freeze the association's assets until the matter was resolved.

While looking for a way to finance the proposed Soldiers and Sailors Memorial Stadium, the fair appropriated funds for construction of a $50,000 Coliseum on the frame of the old Livestock Pavilion. Boasting a large arena with seating for 5,000, the rebuilt structure was heated and ventilated to permit year-round use. Another $10,000 was spent to add a balcony to the Agriculture Building. A citrus show, flower display and horticulture exhibit were planned for this mezzanine floor.

The fair entertained a proposal to bring Hollywood's favorite cowboy, Tom Mix, to open the new livestock facility. But the $30,000 price tag was exorbitant even by 1929 standards, so a world championship rodeo was booked instead. In other moves, a complete circus was substituted for the usual hippodrome acts in front of the grandstand, and an air exposition, featuring famous pilots and the newest planes, was scheduled for a portion of the Automobile Building. With cooperation from the East and West Texas Chambers of Commerce, the fair instituted a statewide band competition designed to attract more than 3,000 participants.

New midway rides included the Dungeon Railroad and Lindy Loop, named after America's most celebrated aviator. Another aviation pioneer and earlier State Fair favorite, 34-year-old Henry Toncray, died that summer, not in a fiery crash as might have been expected, but from blood poisoning as a result of a mosquito bite.

The stock market fluctuated wildly in September, but experts explained this as a temporary corrective adjustment.

American's rodeo queen, Tad Lucas, appeared before grandstand crowds at the 1929 State Fair.

Considered by many to be the year's outstanding exhibit, the S.S. Pomelo, representing the Lower Rio Grande Valley, was constructed of grapefruit, oranges and lemons with limes used for the smokestacks and kumquats for the lifeboats.

T. E. Jackson, his directors and staff finished preparations for what they were calling "Light's Golden Jubilee Fair." In tribute to the 50th anniversary of Thomas Edison's revolutionary invention, the incandescent electric lamp, the park's main plaza, entry arch and memorial fountain were transformed into a fairyland by thousands of jewel-colored lights. As a centerpiece, mounted on the fair's radio tower, the largest arc light ever seen in Texas glowed in three million candle power splendor.

Another monumental invention had become so firmly entrenched in American life that the 1929 radio show could look at its brief history with nostalgia. Marconi's first wireless set was part of the display.

The state universities at Austin, Texas, and Norman, Oklahoma, resumed their neutral site football rivalry after a six-year layoff. The game sold out. Those who didn't have tickets were encouraged to wait until next year. Mercantile Bank's president, R. L. Thornton, had suggested that a new stadium could be funded by the sale of bonds, and the city agreed to wave further auditorium payments if the fair could manage the financing.

Except for a few rainy days, crowds lived up to the directors' optimistic projections. As a safety measure and to ease the flow of pedestrian traffic, street parking inside the grounds was eliminated. There were no accidents of consequence, but in one spectacular incident, the hydrogen-filled Goodyear Balloon, which had been tethered about 200′ in the air just outside the main gate, exploded and fell on an electrical line.

All power to the park was cut off for 30 minutes leaving dozens of midway riders perched or suspended until repairs could be made.

The 1929 exposition closed with solid attendance and revenue figures. Two days later, on October 29, the bottom dropped out of the market, and stock prices plunged. The boom was over.

1930

The stock market crash did not produce immediate economic chaos, but by spring, business activity was slowing down. Supplies greatly exceeded demand, thereby setting up a chain reaction. As inventories backed up, companies laid off employees. Unemployed workers stopped buying manufactured goods and farm products. Overextended banks failed — more than 1,300 closed in 1930. Deposit losses further reduced spending. More plants shut down putting more people out of work, and the cycle continued.

The State Fair took another look at the proposed athletic facility and cut costs and seating capacity. The unwieldy name was simplified to Fair Park Stadium

Banker Thornton was elected to the board of directors, and his presence appreciably strengthened the organization's financial clout. The fair erected a $60,000 Dairy Building near the Livestock Coliseum, and construction workers were hired to build a $30,000 housing for the spectacular "Battle of Gettysburg."

The "Battle of Gettysburg," a giant historical cyclorama, had been a major attraction at international fairs since the turn of the century.

This giant cyclorama, created by French artist Paul Philippoteaux and 16 assistants, contained 80,000 distinct characters. Seen by visitors at international fairs in London, Buffalo, St. Louis, San Francisco and, most recently, at the 1926 Philadelphia Sesquicentennial, the painting was booked for the 1933 Century of Progress Exposition, but would stay in Dallas until it was shipped to Chicago.

Securing the "Battle of Gettysburg" was only one part of the State Fair's aggressive response to the onset of hard times. Visitors would be treated to a scaled display of famous Navy battleships; an exhibit featuring the dogs, sleds, clothing and other accoutrements used by Admiral Byrd during his much publicized exploration at the South Pole; a mechanized model showing the proposed Trinity River Canal and related industrial district; plus demonstrations by a Chicago meat cutting specialist who wore no gloves or apron while scientifically butchering with a very small knife. Special appearances were scheduled for Jim White, the adventurer who discovered

Groundbreaking for Fair Park Stadium in the spring of 1930. From left: State Fair Secretary Roy Rupard; City Parks officials Foster Jacoby, Mrs. Sansom Smith, and Edgar Hurst; Mayor Waddy Tate shaking hands with Phil T. Prather, State Fair Stadium Committee chairman; engineer Carl Forrest; W.A. Gifford of Gifford-Hill Excavators; and architect Mark Lemmon.

Carlsbad Caverns; Uncle John Hickman, a 110-year-old Negro with memories of the men who fought at San Jacinto and Goliad; and "Old Rip," the horned frog found alive in the cornerstone of the Eastland County Courthouse after 31 years' imprisonment.

Other new sights would include the first use of film in an exhibit, as Ford Motor Company showed movies of cars on the production line, and a replica of the WFAA radio station carved out of 8,400 pounds of Ivory soap by a 16-year-old Dallas boy.

But the 1930 exposition seemed "snake bit" from the start. Management had contracted for the world's champion parachute jumper, but he was killed that summer. Forty rodeo broncs trampled each other to death in a railroad car en route to the fair. And Jack Donahue, star of the auditorium show, "Sons o' Guns," died 10 days before the opening.

When it rains, it pours — and of course it did that, literally, on all the crucial revenue days. Governor Dan Moody wore a suit of Texas cotton to address a "Made in Texas Day" gathering. His teeth chattered as he proclaimed, "Texas clothes are never out of season," and he slipped on an overcoat immediately after the speech.

Spotty attendance at "Broadway Flashes of 1930" forced management to drop all grandstand admission charges. Cave explorer Jim White was assaulted and robbed behind the Exposition Building, and a bloody fight broke out one evening before the rodeo performance. After police clubbed two trouble-making cowboys, their fellow riders and ropers opened the chutes and let steers stampede the arena.

Construction of Fair Park Stadium began in May of 1930. The old wooden structure was torn down, and the new facility was built in the center of the fairgrounds on the racetrack infield.

The playing surface was located 18′ below ground, and a 46′ embankment was built up from the field. Concrete and redwood were used to construct seating for 46,200 spectators. The outside slopes were landscaped, and the bowl was surrounded by 60 acres of parking. It was the largest stadium in the south and designed to permit expansion above the 46 existing rows of wooden benches.

Complaints were voiced about county employees using their vacations to work part-time at the fair. Many felt these jobs should have gone to the growing numbers of unemployed walking Dallas streets. Some of those who had worked were also unhappy. None of the contractors hired to install the "Battle of Gettysburg" had been paid by the promoter.

Undercover officers were assigned to break up an alleged liquor ring operating inside the park after six fairgoers were hospitalized from drinking toxic spirits. But police could only turn up a few hollow dolls selling for 50 cents and presumably intended as flasks.

"Sons o' Guns" received good notices. Jack Haley, who later achieved film immortality as the Tin Woodsman in "The Wizard of Oz," stepped into Donahue's part. The cast also included William Frawley, who would cap a long career by playing grumpy Fred Mertz in the popular TV sitcom of the 1950s, "I Love Lucy."

"Amos and Andy" was so popular in 1930 that WFAA-Radio opened its Fair Park broadcast studio to the public every night so that no one had to miss a single episode.

The new Fair Park Stadium hosted a full slate of college and high school games during the fair. Finally, Texas fans could buy all the tickets they wanted. About 25,000 watched the Longhorns battle the Sooners. A slightly larger crowd showed up for the SMU-Indiana game the following week.

The final count revealed a 30% drop in attendance and a corresponding loss of revenue. By pushing ahead with major construction in 1930 and standing by its commitment to quality entertainment, the State Fair of Texas had demonstrated considerable courage in the face of a crumbling economy. But the handwriting was now on the wall and in the ledger books. To survive the Depression, the fair would have to slash expenses, and along with other businesses across the country, do everything possible to ride out the storm.

1931

In a special election, voters approved 39 amendments to the city charter giving the 260,000 residents of Dallas a council/manager form of government. John Edy was appointed as the first Dallas City Manager in 1931, and Edwin J. Kiest, after 27 years as a State Fair director, resigned to become president of the Park Board.

Money was scarce at every level of state and local government, but the Texas legislature gave a needed boost to the flickering dream of a 1936 celebration honoring a century

Loading the Lightnin' coaster on Dallas Day.

of Texas independence. A 21-member Centennial Committee was created to develop support for the event. Exposition enthusiasts had been promoting the idea for several years, and this was viewed as an important step toward securing the massive funding that would be required.

The fair's directors authorized President T. E. Jackson to appoint a committee to work with state officials in selecting a site and making other plans for the proposed Centennial, but the board's main concerns were holding down costs and finding exhibitors for the 1931 fair. Even with a 20% reduction on the charge of space, many firms simply did not have the money to participate.

Ticket prices for the auditorium show, "Three Little Girls," were rolled back from a $4 top to $2.50. The old grandstand was torn down, and a 22-act combined rodeo and circus, which admittedly was not first class in either category, was booked into the Livestock Coliseum arena.

Even with reduced circumstances, the fair still provided powerful stimuli for the senses and imagination. Visitors in 1931 saw demonstrations of shatterproof glass at the automobile show. They ran their fingers over a fortified limousine, reputedly once the personal car of mobster Al Capone, and felt the baking heat generated by sun lamps at the Dallas Power and Light exhibit. Fairgoers found Ubangi pygmies on the midway and Indian war bonnets in a small natural history museum inside the Agriculture Building. A Psychograph, described as an automatic mind machine and looking like the cumbersome apparatus used to give permanent waves, was on display. The invention lost some credibility when a well-known local historian, Dr. Herbert Gambrell, volunteered to have the wires attached to his head. The machine concluded that the SMU professor had only mediocre ability for remembering dates.

Of course not everyone could be satisfied. One youngster looking for the fat lady was told by a sideshow manager, "Times are hard, Sonny, and fat ladies eat too much, so we ain't got one this year. We got four midget shows instead."

The magical million attendance mark was never mentioned. Fair officials quietly expressed hope that the figure would be higher than the previous year. It was, just barely, but total income was down 25%. There was some consolation for finishing in the black when many state fairs had recorded losses in the $50,000-$75,000 range. But after the exposition closed, staff salaries were cut.

1932

In the spring of 1932, the nation recoiled in horror when the 22-month-old son of its favorite hero, Charles Lindbergh, was kidnapped and murdered.

World War I veterans marched on Washington in a futile effort to secure early bonus payments. Those who refused to return home were eventually routed from their makeshift camp by federal troops under the command of General Douglas MacArthur and Major Dwight D. Eisenhower.

New phrases popped up in ordinary conversation: bread line, soup kitchen, dust bowl, migrant, hobo jungle and Hooverville. The president struggled to halt the skid, still resisting direct aid to individuals, the "dole" which he considered both abhorrent and unconstitutional.

The Dallas Park Board chose not to reopen Fair Park's municipal swimming pool for the 1932 season. It was expensive to operate and difficult to clean, and some members felt it should be demolished in keeping with cutbacks in salaries, personnel and insurance coverage.

Otto Herold, State Fair President, 1932-1933 and 1935-1938.

Otto Herold, one of the key figures in the history of Dallas hotels and founder of the Oriental cleaning and laundry businesses, was elected president of the State Fair of Texas. Herold appointed a 19-member Junior Board of Directors in order to involve younger men in leadership roles. He also instituted further cost-cutting measures. Instead of an expensive Broadway musical, the fair booked a vaudeville showed called "Dream Girl Follies" as its featured entertainment. Ticket prices dropped from $2.50 to a more affordable $1.50.

The drab grey concrete of the front entrance was painted white, and seven other buildings were refinished in white stucco to give the park a new look. "Festival of Light" was chosen as a theme to celebrate the 50th anniversary of electric light in Dallas. Special honor was paid to the three pioneers

A coat of white paint brightened the main entry in 1932.

who established the first light plant in 1882: Alex Sanger, W. C. Connor and Jules Schneider — men who had also served at the helm of the fair. The park lighting system was color-keyed — amber lights directed patrons to the auditorium, green pointed the way to livestock activities, white identified the quickest route for stadium activities, and red lights marked the path to the midway.

Fair officials persuaded the railroads to reduce their rates to a pre-war level of ½ cent per mile on days of major events. They leased 80,000 square feet of outdoor space to Ford Motor Company to build a test ground with miniature hill climbs and other road obstacles. This decision precipitated protest from the Automotive Trades Association which wanted to limit all automobile exhibits to the show it sponsored.

For the first time in three years, the city decorated downtown streets to create a festive fairtime atmosphere. And with the fair's encouragement, Dallas theaters, both stage and screen, attempted to upgrade their programs for the month of October. Traditionally, local entertainment businesses had run up a white flag and scheduled "turkeys" for this period.

There was no formal opening ceremony for the 1932 exposition, but at nine o'clock on Saturday morning, all the iron-lunged whistles of railroads and industry, led by the Adolphus Hotel's giant siren, shrieked for about four minutes.

One of the main attractions was a Seminole Indian Village, which featured native Americans from the swamps of Florida plus a wildlife exhibit and alligator wrestling. The other big draw was Hoot Gibson's rodeo. In between making eight western films each year, the tinseltown cowboy produced rodeos on his California ranch. Now he was taking his show

on the road, an extravaganza with movieland cowboys and cowgirls and outlaw horses captured on the plains of Wyoming and Idaho.

Fairgoers got a close-up look at television. The new wonder had been demonstrated briefly in Dallas during the summer months. There was also the opportunity to see, and perhaps ride in, the Goodyear Blimp. For would-be passengers in the giant sausage, bus service ran from the aviation show in the Automobile Building to Love Field. One large exhibit showcased East Texas' new money crop, tobacco, and another popular display promoted the repeal of prohibition.

Major Jimmy Doolittle served as judge for a model airplane contest which offered a 500-mile plane trip for first prize. Pigeon races provided an alternate form of aerial competition. With starting points in either Texarkana or Mt. Pleasant, the birds flew to a finish line inside the park.

The fair operated with no more than the usual number of misadventures and mishaps. Seminole Chief Tom Billy took ill and died, throwing the village into a frenzy of mourning. An all-night ceremony was followed by burial rites at dawn in an Oak Cliff cemetery. Sultan, the lion star of the motordome show on the midway, also failed to finish out the fair. After first attacking his female trainer and later digging his claws into a 14-year-old spectator, Sultan was given early retirement.

Visitors enjoyed evening fashion shows sponsored by Dallas merchants. The "in" material for 1932 was velvet, an indication of the prevalent desire to forget about the Depression and forego practicality.

It was the first rainless, cloudless State Fair in anyone's memory. Attendance ran about 70,000 ahead of the previous year, but total income dropped again, and the fair was not able to settle 1932's obligations.

Meanwhile, the Centennial was gathering momentum. In November, Texans approved an amendment to the state constitution which permitted the legislature to appropriate funds for the support and maintenance of this one-time exposition. Voters also approved a change in national leadership by endorsing the "New Deal" and Franklin Delano Roosevelt.

1933

Roosevelt advocated a reappraisal of government's role: "The country needs bold, persistent experimentation. It is common sense to take a method and try it. If it fails, admit it frankly and try another."

With industrial output slowed to half its previous capacity, approximately 25% of the labor force unemployed and the nation's banking system near collapse, the 51-year-old president took office. Emergency legislation passed during the administration's first 100 days marked a turning point. Fireside chats reassured the country, and the "alphabet soup" agencies such as the NRA, TVA and FDIC sparked hope of recovery.

Social change was also in the wind. The ratification process necessary to repeal prohibition was underway. Betting on horse races was approved in nine states, including Texas. And the first world's fair since before the war, Chicago's Century of Progress Exposition, opened on a six-mile strip along Lake Michigan. Exhibits emphasized the union of science and industry, and despite the economic turmoil, this fair, which ran for two summers, paid off its underwriters and even yielded a surplus.

The New Deal farm program restricted production and provided subsidies in an effort to raise agricultural prices to parity level with industry. To implement these measures, farmers were paid to destroy existing crops. Ten million acres of cotton were plowed up. Replacement crops for these fields and other approaches to cost-efficient farming were highlights of the agriculture show at the 1933 State Fair of Texas.

The fair's financial problems had been eased somewhat by selling membership certificates, which actually amounted to discounted season tickets good for the next five years. With this money, the fair was able to provide new dormitory space for 3,000 youngsters from 4-H and FFA clubs across Texas.

Management stressed the importance of educational features at what was styled the "Recovery Fair." An evening was reserved for a New Deal rally with speeches by Vice President John Nance Garner and Postmaster General James Farley, but the exposition's primary purpose was still entertainment.

Music lovers applauded a return to the policy of presenting operettas in the auditorium. Three shows were booked over the 16 days: Noel Coward's "Bittersweet," "Nina Rosa" and "Floradora." For sports fans, in addition to 13 football games, boxing was scheduled, and a 1/5 mile flat track was built to accommodate motorcycle events. The legendary Barney Oldfield agreed to race a tractor against the clock over a course beginning in Fort Worth and ending on the fairgrounds.

Beer was sold on the midway for the first time since 1915. Stamp collecting, a hobby popularized by President Roosevelt, was spotlighted in a display that featured over 12,000 different stamps and covers. The police department's exhibit included the pistol used by desperado Harvey Bailey in his daring Labor Day escape from the Dallas County jail. Crime

Fairtime rodeo incorporated a new feature for 1933 — bullfighting. Of course, killing the animal was illegal, but the promoters promised a spectacle with all the lacy trimmings of Madrid. The brave Spaniards entered the arena wrapped in traditional capes, but carrying no swords; the bulls, seeming to sense that the odds had switched in their favor, gored and trampled two of the toreadors during the first performance. Similar incidents occurred every night. The Dallas Humane Society protested the event, but their charges were blunted by the *Dallas Morning News*, which dryly observed that it wasn't the animals that were getting hurt.

concerned and fascinated fairgoers. The trial of George "Machine Gun" Kelly was in progress in Oklahoma City, and the fabled duo of Bonnie Parker and Clyde Barrow had been robbing banks in and around Dallas for the past year.

Negro Day at the fair, an official calendar event since 1930, was widely promoted as "Octeenth." Carnival Night once again offered cash prizes for the best costumes, although contestants were warned that wearing no clothes at all would be grounds for disqualification.

Attendance was up by about 100,000, but the fair's financial condition remained precarious. Total income was only half of what it had been in 1929.

1934

Needing money to stay current with obligations, the directors turned to a logical and familiar source of revenue — horse racing.

Rosser J. Coke, State Fair President, 1934.

Without funds to build a track, the fair, led by its new president, attorney Rosser Coke, entered into a partnership with private businessman R. B. George to construct a complex that would include a grandstand, bleachers, club house, barns and track for roughly $150,000. Receipts from the operation would be shared by the fair and George on a 60:40 basis over the next ten years.

The site selected for the track was in the northeast corner of Fair Park where the state fish hatcheries were located. To obtain this property, the City of Dallas arranged a trade that gave the state a fish hatchery at White Rock Lake.

While preparing to get back into the horse racing business, the directors were also taking a hard look at the state's proposed Centennial Exposition and their own desire for a Golden Jubilee in 1936 to commemorate 50 years of State Fair history. At a board meeting in March 1934, a committee recommended that the fair proceed with its own plans because: ". . . the bill passed by the last session of the legislature very appropriately provides for celebrations at San Antonio on March 2nd; at Houston on April 21st; at Goliad, Brenham, Nacogdoches, Huntsville, and other places identified with early Texas history. And while the bill provides that the Commission shall designate one city as the place for the central exposition, it appears evident that likely there will be but little difference in the patriotic appeal and the importance of many separate celebrations which will be held in various historical spots of the state."

That there proved to be a significant difference among the separate celebrations, and that Dallas was named the central exposition site over cities with greater historical claim, was due largely to the foresight and perseverance of one man, Robert L. Thornton.

When the Texas Centennial Commission finally defined its requirements for a host city and set a September 1 deadline for presentations, Thornton grabbed the initiative. Persuading and cajoling, strong-arming and convincing others to help, he put together a Dallas proposal. The final version, endorsed by the entire State Fair Board at Thornton's request, offered the Fair Park plant plus additional acreage; $3.5 million in funding to be raised by municipal bonds; and $2 million underwritten by local businessmen. The bid, with a total value of $9.5 million, almost doubled the nearest offer, and Dallas won the right to host the central celebration.

For the leadership of the State Fair, there was a bittersweet edge to the triumph. Though several directors would take active roles in the Centennial management, though Centennial funding would dramatically alter the form and function of Fair Park, and though financial compensation was part of the agreement, the State Fair of Texas would in effect go out of business for the next three years and relinquish its property and place in the hearts of Texans to a separate and independent organization. The fair would celebrate its own birthday belatedly in 1938.

But for the moment, there was 1934 to consider. Most Americans were engaged in valiant, but anonymous struggles to pay bills and keep businesses open. A few labored in the spotlight. Cartoonist Harold Gray earned $100,000 per year for drawing the adventures of Little Orphan Annie. Shirley Temple made her film debut in "Stand Up and Cheer" and quickly became the nation's number one box office star. J. Edgar Hoover enjoyed a meteoric rise to prominence as the FBI outwitted and outgunned John Dillinger, Baby Face Nelson and Pretty Boy Floyd.

The fair opted for variety in 1934. A multi-faceted entertainment package featured Bicycle Day — free admission for cyclists, an egg-laying contest, a pond full of

Horse racing returned to the State Fair agenda in 1934. A new track and grandstand were built on the site previously occupied by the state fish hatcheries.

bullfrogs imported from Louisiana swamps to demonstrate the viability of frog farming in Texas, and Bob Roberts, billed as the only living man who could swallow a regular Ford automobile axle. "The Show of A Century" offered gags, leggy dancers and even nudity on the auditorium stage.

Pro football was introduced. The hometown Dallas Rams met the Memphis Tigers in an American Professional League game played on the first Sunday of the fair. Sports buffs also watched the city's first marathon race run over a 26 mile course through downtown and the suburbs on Dallas Day. Horse racing proved moderately successful, though not on the same scale as the better outfitted track at Arlington Downs.

The Marsalis Park Zoo created a one-of-a-kind exhibit. The zoo had an old elephant named Wilbur, a longtime favorite of Dallas school children. That fall, without warning, the ancient pachyderm went mad and died. The zoo put Wilbur's stuffed head on display at the fair, and youngsters stopped to say goodbye.

Southwest Conference teams ran up an impressive string of victories during the fair. In football, SMU rolled over Oklahoma A&M, and Texas shut out Oklahoma. In the arena of animal husbandry, Texas A&M won the egg-laying championship with a team of white leghorns.

Attendance crept back above the 900,000 mark, and revenues were up — if only slightly. There was a certain lighthearted frivolity that had been absent the past four years. The 1934 State Fair of Texas primed public interest in the big show ahead.

The Texas Centennial

1935-1937

1935

"The Central Exposition must be Texanic in its proportion . . . continental in its ideals . . . as big and great and beautiful and inspiring as is humanly possible . . ." according to one preliminary report.

"THINK . . . TALK . . . WRITE . . . TEXAS CENTENNIAL IN 1936," urged members of the publicity committee.

And from a promotional flyer: "The Centennial is the biggest job proposed for Texans since wrestling liberty from foreign despotism."

Centennial supporters occasionally got swept away by their own enthusiasm and hyperbole, but it was impossible to overstate the scope and complexity of what had to be accomplished in little more than 18 months.

The first step was to create an organization to run the central exposition. Up to this time, the structural framework of the Centennial consisted of committees which begat boards which begat commissions which begat more committees, with local, state and federal involvement, even as funding was anticipated at all three levels. The new decision-making body, the Texas Centennial Central Exposition Corporation, was headed by the three giants of Dallas banking: Fred Florence of Republic, R. L. Thornton of Mercantile and Nathan Adams of First National. The corporation chose an administrative staff which included former Wichita Falls mayor Walter Cline as managing director; State Fair president Otto Herold, assistant director; retiring Dallas mayor Charles Turner, financial director; Ray A. Foley, director of works; and William A. Webb, purchasing agent. Architect George Dahl was named technical director, and he assembled a team of over 100 architects, artists and craftsmen. Donald Nelson, who had worked on the 1933 World's Fair in Chicago, was hired as chief designer.

A team of sculptors headed by Raoul Josset and Lawrence Tenney Stevens was assigned the responsibility for creating a collection of heroic statuary for the Texas Centennial Exposition.

The National Cash Register Company constructed a 65′-tall replica of its signature product as an exhibit pavilion.

Months passed before all the pieces of the financial puzzle were in place, but ultimately the $9.5 million pledged in the initial Dallas bid was augmented by a $1.2 million grant from the Texas legislature to build the Hall of State, another $500,000 from the same source for advertising, $1.6 million in federal appropriations and $500,000 from municipal bonds previously approved for construction of an art museum. The remaining costs were paid by private individuals and companies.

Backstage maneuvers on the local political front assured an atmosphere that would be attractive to visitors. Hardline city manager John Edy was replaced by the more flexible Hal Mosely, and a tacit understanding existed that city leaders would overlook prostitution, gambling and strip shows during the run of the exposition.

Finally, an agreement was reached regarding the State Fair's role in the Centennial. Under the terms of this contract, from mid-1935 through 1936 and with an option for 1937, the fair gave up all rights relating to its existence in Fair Park, including any voice in determining physical changes to park properties and any revenues from park operations. In return, the Centennial Exposition Corporation agreed to assume the fair's obligations for stadium bond payments, current accounts, insurance and certain other liabilities during this period. A supplementary contract provided $15,000 for the use and occupancy of the fair's premises and specified that the corporation would, as far as appropriate, absorb fair operating personnel into its own organization and employ the fair's executive secretary, Roy Rupard, in a position that would allow him to keep an eye on State Fair interests.

The fair's offices ceased to exist when the front gate structure was torn down. The Agriculture Building was renovated to serve as the administrative headquarters for the Centennial staff. By the summer of 1935, most of the facilities

As part of George Dahl's master design, the Gulf Clouds Fountain was moved from the main plaza to a new location in front of the auditorium.

At the height of the Depression, the Texas Centennial offered employment to thousands of laborers.

on the old fairgrounds had been demolished or stripped to their core in preparation for extensive remodeling. All that remained were portions of the Agriculture and Exposition Buildings, the Livestock Coliseum arena, the Automobile Building, auditorium, swimming pool, grandstand and the stadium.

A new Fair Park was built by an army of more than 8,000 laborers in less than one year at an estimated cost of $25 million. The physical plant consisted of more than 50 structures, waterways, massive pylons, terraces, sculptures and murals expressed in an innovative architectural styling later described as Art Deco, Texas Colonial, Southwestern Classical or simply, according to George Dahl, Texanic.

After prolonged negotiation with nearby property owners, the City of Dallas acquired 26 acres to complete the southwestern quadrant of the park. This property was called the "civic center" and designated for construction of a lagoon, Band Shell and five museums.

As the months wore on and pressures increased, personality conflicts undercut Cline's ability to direct the operation. He resigned in September of 1935 and was succeeded by Otto Herold, who subsequently resigned about a month later. Finally, William A. Webb was appointed managing director, and the executive committee was realigned. The death of financial director Charles Turner further complicated the arduous task of administering the project, but on June 6, 1936, the show and most of the showplace were ready for the official announcement by U.S. Commerce Secretary Daniel Roper, "America, here is Texas."

1936

(top right and below) Opening Day activities included a downtown parade viewed by 150,000 spectators and a gala "Ceremony of Flags" in the Cotton Bowl. Governor James Allred opened the front gate with a $50,000 key crafted for the occasion by jeweler Arthur Everts. The key disappeared from a showcase in Everts' store in 1952 and was never seen again. (top left) President Franklin D. Roosevelt visited the Centennial on June 12, 1936. (left) The exposition's hostesses were called Rangerettes after the legendary Texas lawkeeping force.

(top) Gulf Oil built a complete broadcast studio that allowed visitors to watch all phases of radio production. Art Linkletter got his start in broadcasting as the station's program manager. (left) Ice skaters entertained on an outdoor rink in the Black Forest Village, a complex of small shops and cafes. (above) The Chuckwagon served traditional western fare from a stand modeled after a prairie schooner. (below) Visitors on the midway in 1936.

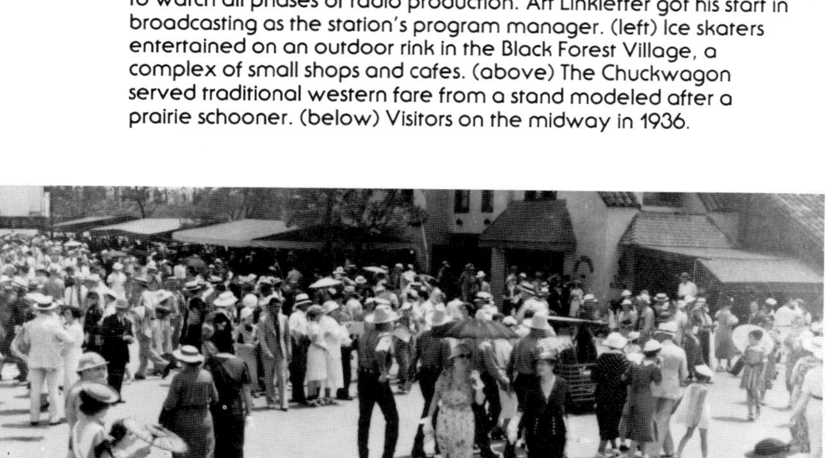

Tourist agency figures indicated that at least one million exposition guests were from outside the state of Texas. The largest attendance during the six-month run, 117,625, was recorded on opening day. (center) Two gargantuan dinosaurs, 70′-long creations by Sinclair Oil, proved to be among the Centennial's more memorable attractions. (below) The main entrance.

(top) Entertainment in the Band Shell ranged from the Salt Lake City Tabernacle Choir to Duke Ellington's Orchestra. (left) The Magnolia Lounge featured a 120-seat auditorium with an observation deck on the roof. (below) The Agriculture Building. (right) The Cavalcade of Texas was presented on an enormous stage with the true-to-life props that included horses, longhorn cattle, covered wagons, a stage coach and more than 300 performers.

(top) Funseekers on the midway enjoyed the racing coaster, Ripley's "Believe It or Not" show, Midget City, Admiral Byrd's "Little America" camp and Tony Sarg's toadstool-shaped Marionette Theater. Private clubs on the Streets of Paris offered adult entertainment, mixed drinks and an atmosphere of sinful sophistication. (left) Centennial guests during the month of August included Vice President John Nance Garner and the Sons of the Pioneers, a popular singing group that featured Len Slye — later known as Roy Rogers. (below) The Law West of the Pecos was memorialized by a reproduction of Judge Roy Bean's famed saloon/courthouse.

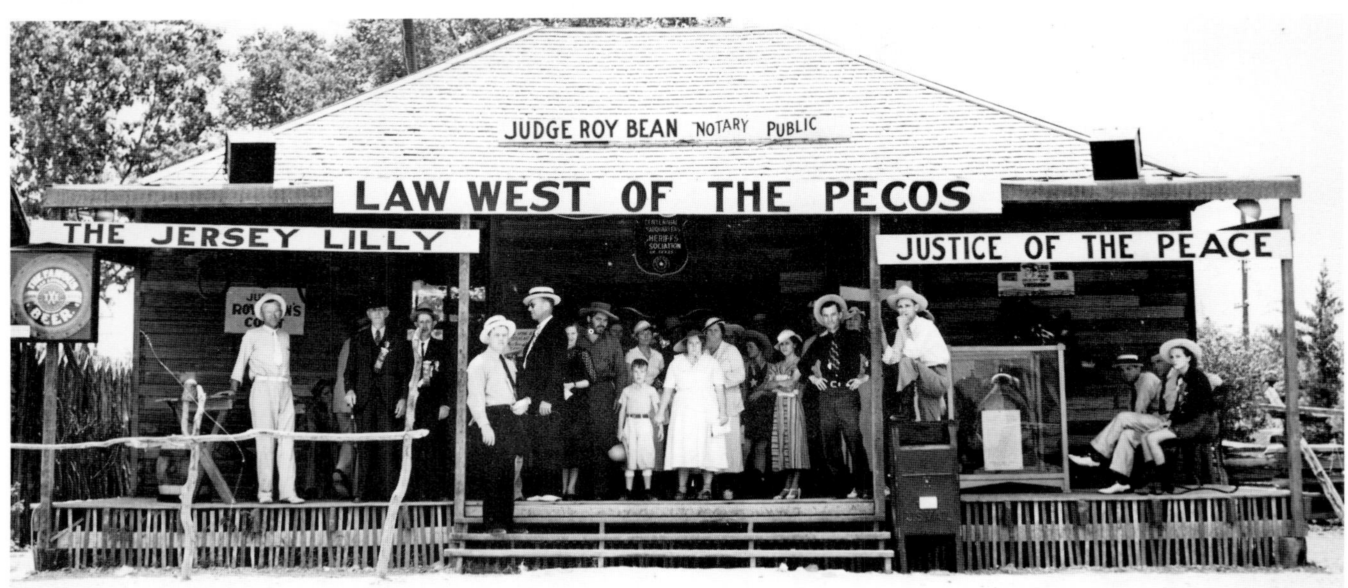

(top) Each evening, the exposition dazzled visitors with lighting effects that splashed a rotating spectrum of color over the buildings, artwork and Esplanade. (left) Native Texan Ginger Rogers made a two-day appearance and crowned the Centennial queen. (right) Grounds transportation was furnished by 15 streamlined coaches, each with a 60-passenger capacity. (below) Exhibits in the Federal Building included a $4 million stamp collection, a giant census board and weather map, a wind tunnel used to test military aircraft, a model of Boulder Dam and curios from the territories of Hawaii and the Philippines. A copy of the United States Constitution was displayed under heavy glass.

(top) The centerpiece of the Centennial, the Texas Hall of State, stood at the head of the Esplanade, a 700′-long reflecting pool. (left) Shakespearean actors performed in the Globe Theater, a replica of a 16th century London playhouse. (right) General Motors renovated Fair Park Auditorium for its exhibit building, installing air conditioning and raising the lower floor to stage level for new car displays. (below) The Cavalcade of Texas drew an audience of 1.2 million over the six months.

(top) "Dallas for Culture — Fort Worth for Fun" was the gist of a promotional campaign created by Broadway showman Billy Rose for Cowtown's rival Frontier Centennial Celebration. Posters and signs advertising the Fort Worth event were prominently displayed all over Dallas. (left) Controversy surrounded the degree of public nudity permitted on the midway. Strip shows displayed the more obvious talents of Corinne the Apple Dancer, Paris Peggy and Lady Godiva. (right) The oriental section of the Street of All Nations. (below) A fountain in the lagoon with the new Fine Arts Museum in the background.

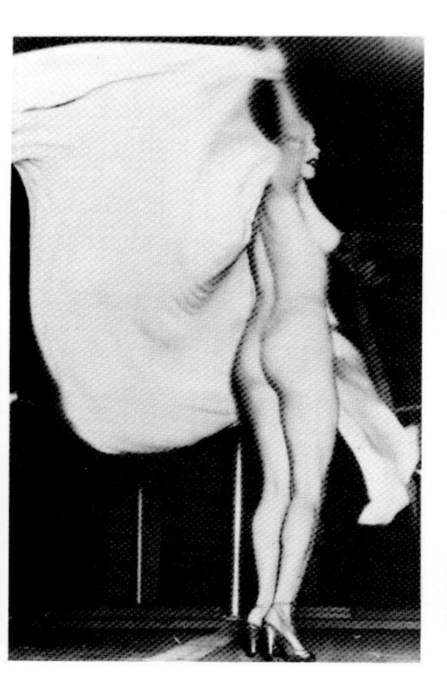

(top left) The Spirit of the Centennial, designed by Raoul Josset and executed by Jose Martin, stood above the entry to the administrative offices. (top right) Food service, Elizabethan dances and curiosity shops were available at the Village on the Green adjacent to the Globe Theater. (left) A performance by the United States Marine Band. (below) Called "the Westminister Abbey of the western world" by former governor Pat Neff, the Hall of State's splendor was enhanced after dark with an aurora of color produced by 24 huge searchlights behind the building.

1937

The Texas Centennial ran for 178 days, from the sunny elation of June and July through a searing August heat wave into the Indian summer ambiance of September and October to conclude under grey November skies. The exposition withstood and perhaps even benefitted from competition provided by Fort Worth's celebration. It overcame the loss of its third managing director, W. A. Webb, who died of a heart attack in early August, and the Centennial finished under the steady hand of Harry Olmsted. In an eventful year, amid news of a Spanish Civil War, Jesse Owens' gold medals, Mussolini's rape of Ethiopia, the Scottsboro Trial and King Edward's abdication for the woman he loved, the Centennial Exposition thrust Texas accomplishments and Texas ambitions into the spotlight.

The official attendance count was 6,353,827. In common with most large scale fairs, the Texas Centennial did not make money. Its success would be measured by the attention it focused on the southwest, by the impact it had in revitalizing the local economy, by the new products and processes — particularly, air conditioning of public facilities — that it introduced into everyday life and by the magnificent plant in Fair Park which was its legacy.

Within days of the Centennial's closing, plans were announced to open another major fair the following June on the same site. Though the Centennial admittedly had been a celebration of Texas victory over Mexico, organizers of the 1937 event expected to obtain the cooperation of Mexico and the nations of South and Central America to produce a five-month Greater Texas and Pan American Exposition. Centennial assets and liabilities were transferred to a new Pan American Corporation, and realtor Frank McNeny, a State Fair Board member, was named director general.

State Fair stockholders met in December, even though the organization would remain on hiatus for another year. The books for 1935 and 1936 showed annual profits in the $50,000

Exposition officials remodeled the Ford Exhibit Building to serve as a Pan American theme pavilion in 1937.

range, the result of compensatory payments and minimal expenses. At this meeting, it was suggested that the directors deviate from formula again in 1938 by producing a 30-day Southwestern Fair in cooperation with neighboring states. No action was taken, but a few days later Lieutenant Governor Woodul proposed a centennial of Texas statehood be celebrated in 1945, further evidence of the high regard accorded to long term expositions at that time.

Less support was given to a venture put together by Dallas businessman Curtis Sanford. Sanford's dream was a post-season intersectional football match-up to rival the New Year's Day game played in Pasadena, California. On January 1, 1937, in the first Cotton Bowl Classic played at Fair Park, TCU edged Marquette before only 12,000 fans, and promoter Sanford had to dig $6,000 out of his own pocket to meet expenses.

The Pan American Exposition was intended to piggyback on the popularity of its predecessor, and without the tremendous initial expense, it was hoped that this second major fair might operate at a profit. But organizers soon ran into problems. Little money was available for the structural and cosmetic changes needed to create the illusion of a new event, and Latin American interest in the project was lukewarm.

Most of 1936's exhibitors and attractions returned for the second season. The Ford Building was converted into the Pan American Palace to house national exhibits, and a craft village was erected to showcase Latin American artisans. The General Motors Exhibit Building became the Pan American Casino, a dining and dancing establishment featuring revue-style entertainment by such name entertainers as Ted Fio Rito's Orchestra and the dance team of Volez and Yolanda. The Cavalcade of Texas was restaged as the Cavalcade of the Americas.

George Preston Marshall, sports entrepreneur, owner of the Washington Redskins and husband of film star Corinne Griffith, was hired as the exposition's entertainment director

(below) Fair Park Auditorium, used by General Motors during the Centennial, was converted into a dining and dancing establishment called the Pan American Casino.

(right) The grand finale of the Cavalcade of the Americas.

at an announced salary of $100,000 for the season. Marshall was responsible for the summer's most innovative event, an Olympian athletic spectacle called the Pan American Games.

The park was decorated in a four-color scheme combining Aztec red, Mayan blue, Toltec green and Incan gold, and 200 palm trees were planted in front of the Hall of State. The Road to Rio replaced the Streets of Paris; official hostesses were rechristened "Texanitas"; and exotic dancers on the midway used sombreros instead of fans.

Amelia Earhart was well into her ill-fated round-the-world flight and the Duke and Duchess of Windsor were honeymooning when the Pan American opened on June 12, 1937. Eight national exhibits were in place in the theme building with others still expected. Marshall had put together an exciting entertainment lineup which included appearances by Jack Benny and Mary Livingstone, Rudy Vallee, Phil Harris and Benny Goodman. The Pan American Games had received AAU sanction. Competition in soccer and track and field was slated for mid-July. Boxing matches were scheduled in August.

Certain aspects of the Pan American exposition were well received, but the event as a whole did not generate the support or enthusiasm expected. The Latin American pavilion was never fully completed. Peru was weeks late setting up, and Uruguay's exhibit didn't even arrive until October. The admission charge to the Cavalcade was eliminated after the first month, and the pageant shut down for good in late September. Responding to pleas by concessionaires, all gate charges were removed during the final month in an effort to stimulate attendance.

In the end, the tangled finances of the Pan American reverted back to the original Centennial Corporation, and settlements were reached on losses. The State Fair of Texas faced sizable start-up costs, but it returned to a remarkably improved property and promptly resurrected its plans for a Golden Jubilee Celebration in 1938.

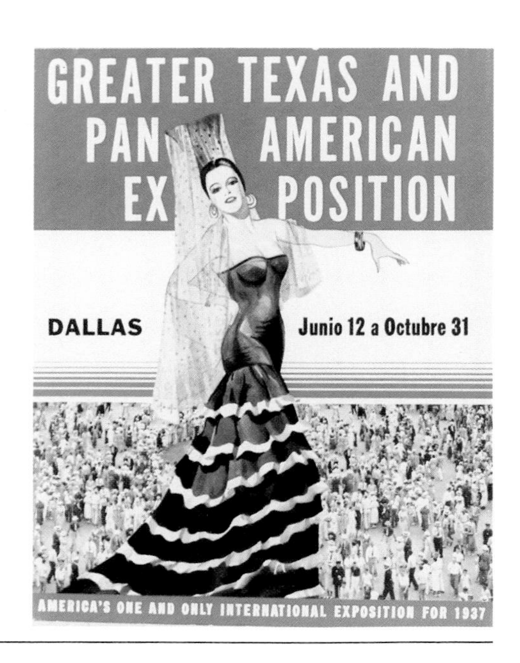

Bigger and Better Than Ever

1938-1945

1938

"Southwest's Exposition To Be Restored After Three Years," announced a *Daily Times Herald* headline. Inside, an editorial cartoon showed the Golden Jubilee edition of the State Fair of Texas pushing war scares off the front page.

Dallas welcomed the fair. If people had become sated with spectacle by the end of the Pan American, they were nevertheless ready for a traditional two-week show in October of 1938. Downtown merchants paid for street decorations which alternated gold bunting with Texas and American flags.

In the years between state fairs, the lingering specter of the Depression had gradually given ground to concerns about the threat of war in Europe. New agencies and new legislation were reshaping the business and social fabric of the country. During this interval, Texans again lost the right to bet on

Fritos were successfully introduced to the public during the Texas Centennial. In 1938, the company's exhibit moved to the newly-designated Castle of Foods.

horse races. The fabled Texas Rangers were assimilated into the Department of Public Safety, and a new figure on the state political scene, radio personality and super flour salesman W. Lee "Pappy" O'Daniel, won the Democratic nomination for governor without a run-off. O'Daniel had first caught the attention of Texas voters with a daily radio program of sacred and hillbilly music which featured the "Light Crust Doughboys."

The Centennial had revived Dallas business. Construction of the Triple Underpass and Trinity River viaducts created new jobs, and petroleum-related industries sprang up to serve the productive East Texas fields. In 1937, R. L. Thornton organized the Citizens Council, a select group of businessmen who would become the power-behind-the-scenes to guide the city's destiny and annoint its leaders over the next four decades.

Fair Park returned to its earlier role as a public recreational facility. The swimming pool reopened, and the city's first lighted softball diamond was completed for summer play. Temporary Centennial structures were razed, and several major buildings, including those that had housed the Ford and petroleum exhibits, the Gulf Broadcast Studio and the Hall of Negro Life, were torn down.

The fair began the year with a staff of eight, two secretaries, $600 cash and a sizable reorganization task. Deferred loan payments came due and wiped out the bank reserves. A salvage operation financed the 1938 fair — Pan American leftovers that couldn't be used were sold. The Federal Hall was remodeled for educational displays; the Electrical and Communications Building became the Castle of Foods; and

As part of the State Fair's 50th anniversary celebration, front pages from Texas newspapers were sealed in an iron crypt inside a statue honoring the exposition's founders.

the auditorium and General Exhibits Building were restored to their original use. Huge neon signs with gold letters identified the six largest buildings, a distinctive touch in keeping with the Jubilee theme.

In front of an opening day audience that included 300 descendants of the exposition's founders, Judge J. J. Eckford, State Fair president in 1912, unveiled a statue honoring these pioneers. During the ceremony, front pages from newspapers across the state were sealed within an iron crypt in the monument, and the key was given to the Texas Press Association to keep until the container would be reopened in 50 years.

New attractions for 1938 included Bozo, the world's only mind-reading dog; a bait and flycasting tournament held on the lagoon; and the first flower show presented by members of Dallas garden clubs. Activities for flower fanciers and dog breeders were scheduled in the old Art Building, constructed in 1908 and showing its age.

The Hall of State reopened officially on October 12, under the management of the Dallas Historical Society, as the fourth permanent museum in Fair Park. The public facilities for art, natural history and aquatic life were already being operated year-round by the city.

High wire acts were popular at the 50th State Fair. Artists and acrobats performed from lofty heights above the Esplanade and in front of the Magnolia Lounge. A third troupe created an aerial ballet on top of the auditorium.

Golden Wedding Day drew 200 couples, and Beauticians Day spotlighted the Cinderella Twins — one with make-up, one without. Dallas Day disappointed everyone. Only 10% of the city's retail businesses closed for the occasion, and attendance trailed far behind projections. Modern merchandising had overtaken this particular fair tradition.

The Texas Hall of State was one of four permanent museums open to fairgoers in 1938.

"Jubilee Follies," a locally-produced, low-budget extravaganza, was the surprise hit of 1938. The best seats in the auditorium were priced at only 75 cents, and more than 56,000 people bought tickets forcing management to hold the attraction over three extra days.

Attendance fell just shy of one million, and while revenues were satisfactory, it was obvious that the post-Centennial State Fair, with an expanded plant and year-round responsibilities, needed a firmer financial base. This was accomplished by the sale of $175,000 in bonds. The restructured fair elected Harry L. Seay, president of Southwestern Life Insurance Company, as its new chief executive and signed another 20-year contract with the City of Dallas.

1939

A scandal involving two members of the Dallas Park Board surfaced in the first month of 1939. Allegations of pay-offs, nepotism, juggling of specifications and sale of city materials for personal gain eventually led to criminal charges.

In April, Dallas voters supported a ticket endorsed by the Citizens Council and elected J. Woodall Rogers as mayor. Youthful James Aston was named city manager, and another promising young city official, 31-year-old Louis B. Houston, was hired as the new park director.

Harry L. Seay, State Fair President, 1939-1944.

In a decade of world's fairs, the most ambitious and expensive international exposition ever attempted opened a two-year run on a 1,200-acre site in Flushing Meadow, New York. Television, robots and nylon stockings were spotlighted in "The World of Tomorrow," and many shows and celebrities from this gargantuan spectacle later appeared at the State Fair of Texas.

During the off season, the Metropolitan Opera made its first spring tour stop at Fair Park. City-owned radio station WRR remodeled the south end of the General Exhibits Building for a broadcast studio. The fair leased the Administration Building to the Farm Security Administration and listened to preliminary inquiries from the war department about the possibility of stationing troops in the park should it again become necessary.

War broke out in Europe on September 1 when Hitler's forces invaded Poland, at last provoking England and France into declared opposition. Most Americans still favored a policy of neutrality. All Polish resistance was crushed in less than a month. With these clouds on the horizon, the State Fair of Texas kicked off what proved to be a record run.

The hobby show offered a unique showcase for collectors and some eye-catching displays for visitors.

As each generation of fair leaders had learned, the one essential element for a successful exposition was the one over which they had no control — good weather. Sunny days and balmy evenings could overcome almost any problem. Conversely, even first-rate talent and costly attractions faltered in cold, wet, windy conditions. Occasionally, luck, finances, know-how and sunshine converged.

Such was the case in 1939. The grounds were overrun with entertainment. Even the livestock department got into the act with Ferdinand, the educated bull from Coryell County who shook hands, rolled barrels and assisted judges. The Great Wallendas worked on bicycles 75′ above the Livestock Plaza, and the biggest attraction was the fair's first National Hereford Show.

Leo, the lion whose roar announced every MGM picture, starred in a small circus presented on the grassy circle in front of the Educational Building. Hopi Indians danced authentic snake dances aided by authentic rattlesnakes. Hobbyists displayed their hoard of matchbook covers, salt and pepper shakers or railroad relics. One collector used 2,000 buttons to build a replica of the Irving Methodist Church. Jitterbug contests flourished on the midway, and the favorite new gastronomical treat was an extra-long hot dog called the Coney Island.

Science and technology earned respectful attention. A water culture show demonstrated methods of growing plants and vegetables without soil. Lone Star Gas displayed a giant world map with neon tubing stretching from Washington to Warsaw, the distance equivalent to laying all of the company's pipes end-to-end. The first-ever chemurgic show illustrated

A huge crowd gathered every day to watch the small circus sponsored by Southern Select Beer.

The sensational "Folies Bergere" came to the fair from the Golden Gate International Exposition in San Francisco, but proved to be a box office disappointment.

how the wonders of chemistry could be applied to agriculture. Fairgoers marveled at pillowcases made of spruce chips and women's clothing woven from milk-based fibers.

"The Folies Bergere" promised and delivered the naughtiness and nudity of a Paris revue. The city welfare director was dispatched to the show to see if censorship was required, but that gentleman prudently decided that such decisions were beyond his authority. The show's star was Mademoiselle Corinne, fondly remembered by Centennial visitors as "Corinne the Apple Dancer." One critic commented that the celebrated stripper's costume had been reduced to an apple leaf.

The automobile show made use of two pretty girls to attract and dazzle crowds. The dog show added obedience trials to its lineup. And by the time the flower show opened, there were 31 clubs entered instead of the original 30 — one group had disagreed about its display and split in two.

More than 70 years after Lee's surrender, a handful of veterans were still able to participate in Confederate Day activities.

"Mr. Dodge," a popular quiz program, was broadcast live from the Cotton Bowl over KRLD. A Thursday evening crowd, estimated at 25,000, showed up to watch and hope for a chance at the grand prize: a new Dodge sedan. Appropriately, the car was won by a Mrs. Fair of Dallas.

Dr. Francis Townsend, proponent of the old age pension plan bearing his name, came to the fair to speak to 1,000 followers at an afternoon meeting. There was a "Garner for President" booth in one building, but the Texas-born vice president did not make an appearance.

Texas and Oklahoma drew the largest crowd in the history of their rivalry — 28,000. Police announced they had uncovered a new form of gambling and were cracking down on betting on football games.

A bell in the tower of the Educational Building tolled each hour, and the turnstiles counted each visitor. At the end of 16 days, 1,036,708 fairgoers had passed through the gates. Attendance would never miss the million mark again.

1940

While Americans lined up at movie theaters to watch epic scenes of horror and destruction in "Gone With the Wind," a modern war raged in Europe. Hitler unleashed his armored divisions against the west, and in a span of less than three months, the Germans defeated and occupied Denmark, Norway, Holland, Belgium and finally France.

The Magnolia Variety Show.

Sentiment regarding neutrality had shifted, and Americans strongly supported aid for the Allies as the Battle of Britain began. The bombing of London was in its fifth week when the 1940 State Fair of Texas opened, and the exposition's emphasis on national defense and patriotism reflected the mood of the country. Uncle Sam's crack motorized division, the Second from Fort Sam Houston, rolled through the grounds on the middle weekend. The division set up its field kitchen and baking oven behind the General Exhibits Building, and the public was given a chance to sample the famous army bread which remained edible for a month. Both the Army and Navy operated recruiting booths, and at a midway archery game fairgoers could vent their feelings by taking aim at clay replicas of the Nazi dictator.

A newcomer to Texas, North American Aviation, which was building a $7.5 million factory at Hensley Field near Grand Prairie, sponsored a large exhibit of aircraft parts and aerial photo murals. The annual chemurgic feature was a National Plastic Show. In the same area, a television studio had been set up with broadcast capability into various buildings on the grounds.

There was no auditorium show. Instead, an outdoor spectacle called "Americana" was staged in front of the grandstand. Spectators paid from 25 cents to one dollar for wholesome, patriotic entertainment performed by a 150-member cast.

Visitors to the Lone Star Gas exhibit were given an opportunity to record their voices and hear the results played back immediately. The building, built by the company as the Centennial Hall of Religion, also featured a glassed-in engine room containing gas-operated air conditioning equipment.

Fans of the Texas and Oklahoma football teams arrived in Dallas the day before the big game. There was cheering and merrymaking downtown — both school bands tied up traffic by marching up Commerce Street into and through the lobbies of the Baker and Adolphus hotels — but no rowdiness was reported. A crowd of 35,000 watched a 19-16 Longhorn victory.

Youngsters took advantage of a Kids' Day special that priced all the midway rides at just a nickel. Another promotion offered free admission to persons over 60 on Townsend Day.

During the last week of the fair, a booth opened in the Educational Building to accommodate any of the 16 million American men between the ages of 21 and 36, who had been ordered to register for the first peacetime draft in this country's history.

Management accepted congratulations for another new attendance record. The fair finalized acquisition of a valuable petroleum exhibit and made plans to convert the old Livestock Coliseum into an ice arena in the coming year.

Winners in the annual Baby Doll Parade on Negro Day in 1940.

1941

"Opera Under the Stars," a series of operettas produced by the Shubert organization for the State Fair of Texas, introduced summer theater to Dallas in 1941. The shows were presented in Fair Park's outdoor amphitheater, now renamed the Casino. Audiences enjoyed "The Chocolate Soldier," "Rose Marie," "The Merry Widow" and nine other one-week productions. By season's end, the operation showed a $600 profit. Though this was more than offset by the fair's $8,000 expenditure for remodeling the Band Shell, there was broad-based support to turn the experiment into an annual event.

The new petroleum exhibit used murals, maps and animated figures to depict the history of oil from 1857 to the present. Built by Petroleum Industries Inc. for the New York World's Fair, this elaborate display was given to the State Fair for the cost of shipping it to Dallas. Under Harry Seay's aggressive leadership, a program of presenting the organization's needs and goals produced numerous donations of property and materials in the early 1940s.

In the renovated arena, Ice Capades, described as a musical extravaganza adapted to a new style theater, was the overwhelming hit of the 1941 fair. Earl Carroll's "Vanities" played in the auditorium marking a return to an indoor show and titilating advertising. Carroll also produced entertainment for the Cafe Esplanade, a cavernous, canopied night club erected inside the old Automobile Building. With very few new car models available for 1942, the automobile show had been cancelled. A Texas girl, Wee Bonnie Baker, who had catapulted to fame singing "Oh, Johnny!" appeared at the cafe with the Orrin Tucker Orchestra.

Fair officials shifted the cake contest to the General Exhibits Building in 1940. Competitive events for women were scattered throughout the grounds.

Rural America had finally claimed a share of the national economic recovery. More livestock filled State Fair barns than ever before. The third consecutive National Hereford Show drew 500 entries. More than 2,000 sheep and goats competed for increased premiums, and poultry show numbers climbed to 7,500.

For the first time since the Cotton Bowl was built in 1930, the 46,000-seat stadium sold out for the Texas-Oklahoma game and in the process created a new problem for the Dallas Police Department — ticket scalping.

The State Fair operated its first independent midway in 1941. Contracting for rides and shows on an individual basis allowed for more selectivity and control than booking an entire carnival. By virtue of its size and late autumn dates, the fair had reached a point where it could choose from the best attractions available rather than accepting a package deal.

Military exhibits dotted the grounds, and talk of war dominated conversations. Newspapers dramatically reported the advance of German troops to the outskirts of Moscow and offered commentary on the militaristic leanings of Japan's new premier, Lieutenant General Hideki Tojo. Texas senator Tom Connally spoke to a gathering of 20,000 for a defense rally in the Cotton Bowl. Army engineers demonstrated how quickly they could build a pontoon bridge across the lagoon, and a sham battle was staged in the stadium featuring an artillery and machine gun attack on a brush-covered hill at one end of the field. As a show business touch, one of the chorus girls from "Vanities" sang "Rose of No Man's Land" to a mock-wounded soldier as he was carried away by stretcher bearers.

The war spawned other fair activities in 1941. RAF pilots, training at Terrell, provided the opposition for an afternoon of Anglo-American soccer. And fairgoers with patriotic fervor and money in their pockets could buy defense bonds and stamps at booths in most buildings.

Despite a substantial amount of rain, the exposition edged toward a new attendance record. Management announced that the last Sunday would be "Million-and-a-Quarter Day," and goal was reached with 2,527 to spare.

1942-45

Although the Japanese bombing of Pearl Harbor on December 7, 1941, brought the United States into formally declared conflict, there was no immediate request to use Fair Park for military purposes. In fact, the government's initial response indicated that the facilities were too small for the camp being considered. The fair, therefore, proceeded with groundwork for a 1942 exposition.

Early on the morning of February 10, a fire, possibly caused by a faulty water heater or gas stove, ignited and burned out of control, totally destroying the 1922 Automobile Building and the 1936 addition along the Esplanade used most recently for food exhibits. It was called Dallas' worst fire in 10 years, and damage estimates ranged from $300,000 to $500,000 for the structures and contents, which included basketball courts and a skating rink. The property was insured for $213,000, but with the growing shortage of building materials, the directors put off any immediate decision about a replacement.

At the end of May, it was clear that transportation restrictions could severely hamper production of a fair. President Seay went to Washington and learned that by fall all rail and bus travel and most automobile use probably would be limited to persons having a part in the war effort.

The 1942 State Fair was cancelled, and the organization reduced its paid staff to four persons. The fair continued to operate its summer midway, and this revenue together with income from the stadium and auditorium covered most of the year's expenses.

Beginning in 1943, the government leased various park buildings which gave the fair sizable monthly rental fees. With careful management, the organization used the money

Maybe it was the name: Brown-Bomber, French Fried Hot Dog, Meal-On-A-Stick, even K-9. Or maybe people just weren't looking for gastronomic adventures in the unsettled summer of 1942.

But when Neil and Carl Fletcher opened their first booth on Fair Park's midway, they had to give the product away — cut up in chunks, batter-fried, speared with toothpicks and set out on the counter in a punch bowl. Samplers became customers, and the Fletcher brothers, a pair of song-and-dance men looking to break into the food concession business, had a hit on their hands.

Fletchers' Original State Fair Corny Dogs have evolved into a symbol of the State Fair of Texas that rivals Big Tex. The all-meat sausages, dipped in a guarded-recipe batter, deep fried in vegetable shortening to a golden brown and slathered with mustard are considered the ultimate "fair food." No one knows exactly how many corny dogs fairgoers have consumed over the years, but that number grows by about 500,000 every October.

Back in the early days of World War II, when there was no State Fair and midway food stands operated only during the summer season, the two brothers decided to commercially adapt a novelty item they remembered from their years traveling the tent show and vaudeville circuits. The product was a hot dog in a cornmeal bun baked in the shape of an ear of corn. The process required 25 minutes, much too long for fast food service, so Carl accepted the challenge of creating a batter that would stick on a wiener without becoming greasy or rock-hard when fried. He experimented with at least 60 variations before finding the right mix, and even that formula was fine-tuned for nine more years until the Fletchers were completely satisfied.

The first corny dogs sold for 15 cents. Today's product costs $1.25 and since 1983 has been available at shopping malls and entertainment venues outside Fair Park. The batter is made with cornmeal, flour, rising elements, ice water, salt, sugar and about six other ingredients blended in a 25-pound Hobart mixer. But though imitations abound, corny dog connoisseurs insist nothing compares to Fletchers' — and some swear that the madcap atmosphere of the State Fair of Texas is part of the secret recipe.

The Starlight Operettas in Fair Park provided popular and inexpensive summer entertainment during World War II.

generated by year-round activities to strengthen its financial position despite the absence of fair revenues.

The directors and staff concentrated their efforts toward making a success of the Starlight Operettas, and in 1943, the operation showed a $35,000 profit. Over ten weeks, 263,000 Dallas residents attended the performances, and a larger, more comfortable outdoor theater was high on the board's wish list for future development. At the conclusion of the 1944 season, Charles R. Meeker, an executive with the Interstate Theater circuit, was hired as managing director of the Operettas.

Through the war years, the fair maintained activities in the park as far as permitted. At the same time, the City of Dallas was giving consideration to Fair Park's future. Various plans were discussed including one that effectively would have turned the park around by relocating the main entry to the back side and extending the boundaries from Pennsylvania to Fitzhugh. Under this proposal, the one-time Administration, old Fine Art, General Exhibits, Educational, Old Mill and Magnolia Lounge buildings would have been torn down, and that entire area made an adjunct to the park's civic center.

As fighting ended in Europe and with U.S. forces advancing in the Pacific, the fair selected a new leader for a new era: R. L. Thornton. A committee headed by John W. Carpenter was already working on funding for a new livestock coliseum, and after the Japanese surrender, Thornton plunged ahead with his plan to rebuild the Automobile Building, repair everything that had been neglected and refurbish wherever necessary.

State Fair director John W. Carpenter, the driving force behind efforts to build a new livestock coliseum.

On With the Show

1946-1959

1946

R. L. Thornton fell in love with the State Fair on his first visit in 1889. After a long summer of picking cotton in the Texas heat, young Bob's father made good on a promise and took his nine-year-old son to the great exposition in Dallas. Of all the wonders he saw that day, Thornton particularly remembered an ice cream on-a-stick confection called Hokey-Pokey and the pail of Concord grapes he purchased for 15 cents and carried home to his mother.

At age 65, Robert Lee Thornton was just getting started.

The tall, rugged, cigar-smoking entrepreneur had borrowed $6,000 in 1916 and parlayed that stake into a multi-million dollar financial institution. Beneath his country boy demeanor and colorful speech, Thornton personified the "can-do" spirit often associated with Dallas.

When he inadvertently coined the word "dydamic," Thornton aptly described himself. Father of the Centennial, founder of the Citizens Council, by the end of the war the silver-haired banker was ready to mobilize his energies for a favorite civic enterprise — the State Fair of Texas.

At his recommendation, a $600,000 bond issue to fund the needs of the postwar fair was offered publicly in January 1946 and sold out completely in less than a month. Then, per Thornton's request, the board created a new top level staff position and hired W. H. Hitzelberger as the fair's first executive vice president and general manager.

Through his involvement with the Chamber of Commerce, Bill Hitzelberger had coordinated Dallas Day activities at the fair since the late 1920s. His interest in the organization deepened as he served first on the fair's junior board and, after 1940, as a full director. This background, Hitzelberger's management experience with the Portland Cement Association and his tireless enthusiasm were well matched to the new job.

During the war, the junior board had merged with the senior board, thereby increasing the number of directors from 24 to 48. To streamline the decision-making process of this larger group, Thornton asked that the president be permitted to appoint vice presidents who would assume responsibility for specific areas. With a strong president, an active executive committee and competent staff leadership, management of the State Fair gradually shifted away from the board as a whole and became the function of key individuals.

The fair had money to spend on a new Automobile Building in 1946, but shortages of materials delayed construction. Price controls and ceilings were being lifted, and in general, the country was making a smooth transition to a peacetime economy. Women welcomed Dior's New Look and longer hemlines after four years of skimpy, restricted fashions. Servicemen returned home and enrolled in college on the G.I. Bill. Home permanents and electric clothes dryers appeared on the market, and Kaiser-Frazer became the first new company to make an American car in two decades.

Plainspoken Harry Truman lived in the White House. Jack Kennedy and Richard Nixon were campaigning for congressional seats, and the Nuremberg Trials had just ended when the State Fair of Texas reopened on October 5, 1946.

Apart from changes dictated by the fire loss, fairgoers found the park much as they had left it. The automobile show, boasting only one 1947 model, was staged under mammoth tents, and the Federal/Educational Building acquired still another identity when exhibits sponsored by food manufacturers were switched to this location. There had been some thought of removing the building's distinctive gold-tipped spire a few months earlier when management received a $10,000 estimate for needed maintenance, but the cost of tearing it down was twice that of fixing it, so the repairs were made.

Another graceful tower was newly topped with a blue neon flame, and Lone Star Gas gave the public a good look at its heralded New Freedom Gas Kitchens which offered automatic features and harmonized colors.

Ice Cycles of 1946, a joint production of Ice Capades and Ice Follies, packed the arena, and the Tommy Dorsey Show, which featured comedian Jackie Gleason and the "glamorous Prima Donna Nuda," Gypsy Rose Lee, did excellent business in the auditorium. But the longest lines were those that stretched for two blocks as fairgoers waited for a chance to meet Elsie the Borden Cow. Elsie, a star of the New York World's Fair, stood calmly in the Old Mill Building at the center of the grounds and batted her long eyelashes at delighted visitors.

The petroleum show occupied a third of the General Exhibits Building. One of its features was a model refinery that demonstrated the processes of cracking, absorption, polymerization, dewaxing and purifying. The home show in the same hall presented furnishings, appliances, draperies, bedding, musical instruments and the Colleen Moore Doll House, a miniature two-story castle with infinite detail that cost nearly $500,000 to assemble.

W.H. Hitzelberger, General Manager, 1946-1949.

Texas fairgoers lost their hearts to Elsie the Borden Cow in 1946.

Kids attending the fair on Rural Youth Day enjoyed a free lunch of hot dogs, grape punch and chips. This practice was discontinued in 1974, when it became obvious that the youngsters preferred to buy pizza, corny dogs and Belgian waffles with their own money.

The fair fed lunch to thousands of youngsters at the first Rural Youth Day gathering. Children's Day drew an all-time, one-day high of 218,075, and attendance on Negro Day climbed to 100,000. With crowds this size, parking became a major problem, not just the lack of spaces, but the prices charged by operators outside the grounds. Visitors grumbled about paying as much as $3.00.

Records tumbled across the boards in 1946. Final attendance was 1,639,986, almost a half million more than 1941, and the fair started its new year with an amazing one million dollars in the bank.

1947

One of Bob Thornton's favorite sayings was, "Keep the dirt flying," and the fair's surplus of cash was soon invested in a massive program of park improvements. Construction began on a modern livestock barn, restrooms for several exhibit buildings and lighting towers on the midway. Roofs were repaired. Workmen installed new sod in the Cotton Bowl, filled and graded the racetrack infield for parking, converted the old jockey club into a picnic pavilion and refurbished the Band Shell in time for the summer season.

Fair Park was a beehive of private enterprise in 1947. Goodyear Tire and Rubber Company leased the 1910 Coliseum/Administration Building to use as a warehouse, and Margo Jones opened her theater-in-the-round in the Gulf Oil Building, formerly known as the Magnolia Lounge. A $100,000 roller coaster called the Comet was built by concessionaire Sammy Bert and became part of the permanent midway.

The fabricated steel required for the new Automobile Building was finally delivered, but construction bids came in more than $100,000 over the architect's estimate causing further delay.

Another Thorntonism: "We think big, and we've got a taste for the best," characterized his efforts to secure Broadway's biggest musical hit, "Annie Get Your Gun," for the 1947 State Fair show. It took a steep $175,000 guarantee to producers Rodgers and Hammerstein, but a contract was signed to bring "Annie" to Dallas as the first stop on its first national company tour with Weatherford's own Mary Martin in the lead. With the announcement, ticket requests poured in from all corners of Texas and Oklahoma.

Time magazine used superlatives to report the opening of the nation's largest annual exposition, though the article characterized Governor Jester's speech as "inflated with standard Texasity." The fair also garnered national media attention when two popular radio programs scheduled live broadcasts from the grounds. Ralph Edwards brought his "Truth or Consequences" troupe to the Band Shell on October 13, and two days later, 10,000 women showed up in the Cotton Bowl hoping to be chosen "Queen for A Day." Emcee Jack Bailey offered prizes that included a trip to Galveston, costume pearls, a fountain pen, four dozen roses, dancing lessons and a lifetime supply of stationery. The winner was told to name her fondest wish, and it would be granted. She asked to have the top fixed on her prewar convertible.

Elsie returned, accompanied by an unnamed calf, and down the street at the Health Museum, a pair of goats attracted large crowds to an atomic energy exhibit. "Adolph" and "Satan" were survivors of the Bikini Island test blast.

There were numerous celebrities among the fairgoers that year. Fleet Admiral Chester W. Nimitz, a Fredricksberg native, spoke to the Veterans Day gathering, and World War II's most decorated hero, Sergeant Audie Murphy of Farmersville, was among those honored. Composer Irving Berlin attended the opening of "Annie Get Your Gun," and his music and lyrics sent audiences away whistling and humming about "Doin' What Comes Naturally." Another show visitor was Mary Martin's teenage son. Young Larry Hagman was learning his way around Dallas in 1947.

The fabulous Duke Ellington played for a dance held in the Cotton Bowl Roller Rink on Senior Negro Achievement Day. Both Mondays of the fair had been set aside for black Texans.

State Fair visitors rode the Comet roller coaster for the first time in 1947.

New activities were added as the fair scheduled two Negro Achievement Days to accommodate the crowds.

Ice Capades was a major fairtime attraction from 1941 until 1967.

A collection of paintings by old masters, on loan from the Metropolitan Museum of Art in New York City and valued at $1.5 million, awed visitors to the Fine Art Museum. Architectural drawings by Frank Lloyd Wright were also exhibited.

Fairgoers were finding it difficult to see the whole show in one day, and many were returning to catch what they might have missed: demonstrations of a one-man automatic hay baler, a bubblegum blowing contest or judging of 150 head of hump-backed cattle at the fair's first organized Brahman Show.

Every ticket had been sold — some for more than face value — when undefeated Texas and undefeated Oklahoma clashed in the Cotton Bowl. Bobby Layne quarterbacked the Longhorns, and Tom Landry lined up at fullback. Texas won handily, 34-14, for their eighth straight victory over the Sooners. A questionable call sparked some bottle throwing by disgruntled OU fans, and players retreated to the middle of the field to avoid being hit. The incident led to a ban on glass containers in the stadium.

Records were short-lived at the postwar fairs, and 1947 replaced 1946 as the all-time attendance and revenue champion. In the afterglow of another exceptional year, it was pointed out that with rampant inflation, labor and material costs were not likely to come down, and the board decided to go ahead with construction of the Automobile Building. Longtime secretary Roy Rupard, whose role had been eclipsed with the appointment of a general manager, was nudged into retirement. S. Bowen Cox became secretary with redefined responsibilities. The directors also voted to take on another major project — enlarging the Cotton Bowl to accommodate football fever in the southwest.

1948

Sportswriters would call it "The House That Doak Built," alluding to SMU's All-American halfback, Doak Walker, but in reality and in common with other Fair Park projects, it was the house or at least the upstairs that Bob Thornton built.

The new upper deck on the west side of the Cotton Bowl, which increased stadium capacity to more than 67,000, was financed by yet another bond sale, this one ingeniously structured to reward bond holders with seat options, thus providing not only funds for expansion, but an inducement to buy tickets in the future.

The fair board approved each expenditure without hesitation, though Thornton himself counseled, "Raising money is easy — paying it back is what's hard."

There was reason for confidence. Business had never been better in Dallas. A suburban housing boom was transforming such quiet towns as Richardson, Garland and Irving into flourishing small cities. The fair acknowledged the broader population base when it redefined Dallas Day as Dallas County Day.

Chance Vought Aircraft relocated its entire operation from Connecticut to Grand Prairie in 1948, a move which presaged an exodus of northern and eastern companies to the Dallas-Fort Worth area.

The United States was enjoying a giddy fling with prosperity. Abroad, however, postwar cooperation had broken down. Soviet dominance of Eastern Europe and communist

Displays of foreign-made products were introduced in the General Exhibits Building in the late 1940s.

aggression in Asia created an atmosphere of heightened tension and suspicion — the so-called Cold War.

On October 9, Dallas mayor Jimmie Temple, Park Board president Ray Hubbard and the State Fair's R. L. Thornton joined Texas governor Beauford Jester to dedicate the $800,000 Automobile, Aviation and Recreation Building. Airplanes, boats, buses and the largest showing of new cars ever seen in the southwest greeted first day fairgoers.

Visitors were faced with a smorgasbord of entertainment choices in 1948. Jimmy Durante and Harry James headlined the auditorium show. Paintings by well-known American artists, including Grant Wood's dour "American Gothic," were on display at the Art Museum. "House Party," starring Art Linkletter, was broadcasting live from the grounds on CBS. Kiddie Town brightened the midway with a collection of new rides for the under-12 set.

The textile, culinary and antique shows found a home in the Horticulture Building across from the Band Shell, and spirited competition for ribbons and awards got underway.

Browsers in the General Exhibits Building stopped for a graphic preview of the proposed Central Boulevard. The display pointed up advantages of a modern multi-lane thoroughfare. Texas Highway Department representatives explained that while traffic on ordinary streets was congested, halting and hazardous, the new expressway would remedy these problems.

The Hall of Foods offered a literal feast for fairgoers. Free samples of beef stew, chicken and biscuits, chili, corn chips, pecans, popcorn, cookies, ice cream and coffee were passed out from booths, and concessionaires complained that all the "freebies" were cutting into their business.

More than 67,000 football fanatics paid for the privilege of watching Oklahoma beat Texas in the remodeled stadium. Those without tickets were still able to view the game if they owned or had access to a television set. WBAP-TV had cameras on the sideline for a first-ever telecast.

Attendance records were toppled daily. The huge crowds squeezed into the 187-acre park during a spell of warm October weather caused several mishaps. One lady suffered a broken arm when her skirt caught on the bumper of a police motorcycle trying to maneuver through the crush. Two women were run down by a large Brahman bull that escaped from the livestock area, circled the Cotton Bowl, charged across the midway and up First Avenue before being recaptured in front of the Natural History Museum. And 14 people were hospitalized with symptoms of food poisoning after spending a day at the fair. Health inspectors discovered 500 pounds of contaminated meat when they checked the food outlets on the grounds.

The Rotor was an immediate hit on the midway.

There was concern for the welfare of performers at the Eskimo exhibit, but the stoic natives of Wales, Alaska, endured a Texas autumn with temperatures in the high 80's — 100 degrees warmer than their homeland.

Final attendance rose to 1,892,327, and fairtime income reached a new high. Expenses were up too, but in December, R. L. Thornton declared that the State Fair of Texas was in the best financial shape of its 63 year history.

1949

The New Year's Day game of 1949 dispelled any remaining doubt about building a bigger Cotton Bowl. Ticket buyers had to be turned away. The stadium had sold out repeatedly that season.

Reasoning that if 67,000 seats were a good investment, 75,000 might be a better one, the State Fair Board voted to construct an upper deck on the east side. Since stadium revenues were already pledged to the earlier bonds, the new addition was financed by selling straight 20-year ticket options at $50 each. The strong response to this fund-raising ploy attested to the popularity of college football in the late 1940s. The pro game was also beginning to build an audience. The upstart All-America Conference had merged with the National Football League, and there was talk of future expansion into the southwest.

With some reluctance, the board approved a venture into the mine field of musical comedy production. Charlie Meeker wanted "High Button Shoes" for the fall attraction and felt he could produce the show himself at a profit. Several directors questioned the wisdom of the fair getting into this high risk side of show business. The Starlight Operettas had lost $50,000 the previous summer, and J. W. Carpenter registered his displeasure at the growing emphasis on expensive entertainment and away from livestock and agriculture activities. Carpenter's dream of a new Coliseum remained unfulfilled. Though bonds had been approved by Dallas voters in 1945, other city projects received higher priority.

Meeker eventually chose the safer route and booked Spike Jones' "Musical Depreciation Review of 1950." Audiences loved its zaniness and lack of sophistication, and the show was a respectable money-maker.

On opening day, curious fairgoers lined up outside the "Man and the Atom" show for a look at life in tomorrow's world. Given assurance that "it is a friendly little atom that greets visitors, not a savage, ever-devouring one," people crowded up to a model showing future industries powered by a nuen (nuclear energy) reactor plant. Other displays brought science into perspective by using such comic strip characters as Dagwood to demonstrate how an atom is split. Questions about the dark side of atomic power were answered in a documentary film presented by the Army and Air Force Recruiting Service. "A Tale of Two Cities" showed the bombing of Hiroshima and Nagasaki.

In a lighter vein, 1949's exposition offered legendary fan dancer Sally Rand, respectably dressed in a costume designed by Dior, in performances every day on the hour from noon until midnight. Midgets from "The Wizard of Oz" also appeared on the midway, and Joie Chitwood's Thrill Show, starring Leo "Suicide" Simon, played in front of the grandstand.

A special exhibit at the Hall of State featured currency from the Republic of Texas and documents of the American Revolution. In the Agriculture Building, Texas A&M took

"Man and the Atom," an exhibit organized by the American Museum of Atomic Energy in Oak Ridge, Tennessee, was one of the highlights of the 1949 State Fair.

5,000 square feet to tell the story of its college system and statewide services. Fairgoers learned about A&M's efforts to mechanize cotton farming in West Texas and save the Texas oyster industry on the Gulf Coast.

Elsie, attended by husband Elmer and son Beauregard, entertained a nonstop parade of guests in her bovine boudoir at the Old Mill. Visitors with a serious interest in beef and dairy cattle, swine, sheep, Angora goats and horses congregated in the barn areas to view a record 2,072 head of livestock from 17 states and Canada. A new breed, an Angus-Hereford cross called Angford, made its debut, and the fair's first all-broiler show drew over a thousand entries.

The outcome was the same, but 7,892 more fans were watching as Darrell Royal led Oklahoma to a second consecutive 20-14 win over Texas. Looking at the full house, fair officials bragged that the Cotton Bowl was now larger than Yankee Stadium.

But the real boasting and celebrating came on the last day when the 1949 State Fair of Texas cracked the two million attendance barrier to conclude a decade of phenomenal growth.

(top) Sally Rand lent her name to shows and sometimes performed in person, but this type of midway entertainment had been sanitized considerably since the "Nude Ranch" days of the 1930s.

(bottom) On October 23, 1949, Mrs. Chester E. Lovelady became the two millionth visitor in a record-breaking year. Mrs. Lovelady, her husband and seven-month old daughter, Rebecca, were welcomed by State Fair president R.L. Thornton.

1950

At the end of World War II, by virtue of its economic strength and atomic arsenal, the United States took on singular responsibility as a world leader. It was not an easy world to lead. In rapid order, communists swallowed up Eastern Europe, conquered China and blockaded the insular city of West Berlin. In 1949, the Soviet Union detonated its first atomic bomb.

A wary America entered the new decade to deal with real and perceived threats to its security. Joseph R. McCarthy, an obscure, but opportunistic senator from Wisconsin, grabbed the spotlight by charging that the state department was infested with traitors. In June, the communist government of North Korea crossed the 38th parallel and invaded the Republic of Korea. A United Nations force, predominantly U.S. troops under the command of General Douglas MacArthur, was sent to the aid of the South Koreans.

More than three million households followed news reports and congressional hearings on television, and their friends and neighbors were buying another 100,000 black-and-white, small-screened sets every week. Top-rated programs starred Arthur Godfrey, Hopalong Cassidy, Howdy Doody and Milton Berle.

James H. Stewart, General Manager, 1950-1965.

Though movie moguls and radio executives worried about competition from the new media, State Fair leaders correctly guessed that the box in the living room would not keep people from attending a once-a-year, two-week special event. Having completed improvements on the Cotton Bowl, the board turned next to 25-year-old Fair Park Auditorium and voted to spend $200,000 to air condition the theater.

Early in the year, general manager Hitzelberger resigned to accept a position with Republic National Bank. He was succeeded by James H. Stewart. Jimmie Stewart, a former football hero at SMU, had played a key role in the Cotton Bowl expansion programs and was serving as secretary of the Southwest Conference when tapped for the fair post.

The 1950 fair was dubbed the Mid-Century Exposition, and "South Pacific," starring Janet Blair and Richard Eastham, returned the excitement of a Broadway production to the now comfortably cooled auditorium.

Other features included "The Drunkard's Daughter," an old time melodrama presented at the new Diamond Garter restaurant on the midway; the nation's highest volume milk producing cow; and big time college football's first double header — a twin bill that pitted SMU against Oklahoma State only hours after the vaunted Texas-OU match-up.

The midway spotlighted a spectacular new thrill ride — the 92' Sky Wheels, built with one giant wheel towering above another and both turning while the entire mechanism revolved.

The fair hosted its inaugural American Saddle Breeders Futurity of Texas Show in 1950 and presented a "Parade of Decades" each evening at 7:45 p.m. This colorful spectacle of floats and bands, the first inside-the-park parade, formed near the Agriculture Building and proceeded on a counter-clockwise route through the grounds.

The atomic energy show shifted to what was now the Science Building, one of the twin flagship structures of the agricultural complex. Displays from the American Museum of Atomic Energy in Oak Ridge, Tennessee, were complemented by exhibits explaining radar, television and the process for making phonograph records. The 33⅓ rpm LP, covered in an eye-catching cardboard jacket, was revolutionizing the business of recorded music. Two booths gave fairgoers hands-on experience with the wonders of science. Those providing a dime from their pockets or purse could have that coin made radioactive, and visitors to the Navy's display were allowed to make simulated bombing runs using a real Norden bombsight, a fascinating exercise that anticipated by 25 years the popularity of video games.

Old Settlers Day featured a husband and wife calling competition and a grandma beauty contest, while Pet Night promised free admission to anyone willing to bring a pet and march in the parade.

Speaker of the U.S. House of Representatives Sam Rayburn, a rancher from Bonham, toured the livestock and grass exhibits. Other celebrities included the fading idol of the war years, Frank Sinatra, who performed before an appreciative crowd in the Cotton Bowl.

Rural Youth Day attracted 75,000 youngsters from farming communities, and a 16-year-old Harris County boy became the first Texan to take the top award in the All-American

Introduced at the fair in 1950, the 92'-tall Sky Wheels incorporated two wheels into one unit for a spectacular new ride.

The women's department featured exhibits of textiles, foods, antiques, potted plants and handcrafts.

The latest models in tractors and farm equipment were displayed on large lots.

Jersey Show. Livestock activities were marred, however, when an elderly man from Durant, Oklahoma, died after being trampled by a frightened dairy cow that broke away while being led to water.

Attendance records were rewritten: 289,307 for a single day and 2,176,519 for the run. President Thornton, who at age 70 was showing no sign of declining vigor or interest, declared with typical modesty that the fair was "as near perfection as capacity and science can make it."

1951

The State Fair of Texas had outdistanced other fairs in the United States in terms of size and scope and trailed only the Canadian National Exposition in Toronto as the largest annual event in North America. And while it was enjoying unprecedented success and financial stability, the continual emphasis on attendance and dollar figures put the organization into the position of having to compete with itself. No fair could be a good fair unless it was better than the last. This spiral was perpetuated by the enthusiasm of its leaders and generally supported by hard numbers, but in the decade ahead, this obsession with records would become a self-inflicted burden.

For its 1951 season, the Starlight Operettas adopted the name "State Fair Musicals" and announced that all performances would begin on time and all shows would be completed, rain or shine, in 68-72 degree comfort. After ten years, the series was leaving the relaxed informality of the Band Shell and the uncertainties of Dallas weather and moving into the air conditioned auditorium.

Fair Park's other summer attraction, the midway, opened for business as usual. The State Fair Board had been urged by

State Fair director Joe C. Thompson, longtime coordinator of Negro Day activities.

director Joe C. Thompson to reconsider the ban that kept Negroes off the rides. Thompson, president of City Ice Delivery, the forerunner of today's 7-11 stores, had been in charge of Negro Day at the fair for many years, but management chose to avoid this sensitive issue. "The time is not right," said general manager Stewart.

"It's a Son of a Gun in '51" was the theme used to promote 16 days of education and entertainment that ranged from "Guys and Dolls" in the auditorium to Aut Swenson's Thrillcade in front of the grandstand. The daredevil show featured ramp-to-ramp jumps, auto ball, motorcycle leap frog and crash roll tournaments.

Visitors studied engines by watching a see-through plastic car running in slow motion at the automobile show. They viewed developments in dairy farming. A milk parlor was set up, and 150 cows were milked under optimal sanitary conditions.

The state's number one beef-producing breed, the Hereford, celebrated 75 years on the Texas range at a Pan American National Hereford Show with $25,000 awarded in premiums. The fair used the occasion of this show to dedicate its new $125,000 open-air livestock judging pavilion. The arena, which adjoined the cattle barns, provided seating for 3,200.

Families wandering through the General Exhibits Building saw the "Range of the Future," which featured an oven capable of cooking a 10 pound roast in 10 minutes and a frozen hot dog bun in five seconds. This wonder would not be marketed for several years, but the public got a close look at other amazing advances in technology: a display of "push button" long distance service, to be available in Dallas shortly, and an exhibit introducing the tower-to-tower microwave radio relay circuit, which would soon bring live network television programs into area homes.

When people had explored all the mind-enriching exhibits, they headed for the midway which offered learning experiences of its own. Curious fairgoers bought tickets to see John Dillinger's crime car. They paid for a front seat at Divena's Water Ballet, where the fully-clothed star descended into a 550 gallon tank and gracefully disrobed. Former heavyweight boxing champion Jess Willard, who appeared at the fair in his prime in 1915, demonstrated fisticuffs in one of the sideshow tents.

No one objected to any of the shows or rides, but a group of citizens complained about midway games. City and fair officials met to determine whether wheels, bingo and other paraphernalia constituted legal games of skill or illegal games of chance. Fair president Thornton closed 62 booths while the question was debated, although a newspaper poll indicated that the public didn't want them shut down. State Representative Doyle Willis of Fort Worth, a leader of the protest,

Winners in the 1951 Trim-a-Hat contest were (from left) Mrs. Don Beck, Stephenville; Miss Joyce Stansbury, Dallas; and Mrs. G.W. Allen, Dallas.

Responding to complaints about gambling in 1951, City Manager Charles Ford led an afternoon sweep through the midway ordering all but one of the games shut down. Fair officials complied. The only booth permitted to remain open was the "African Dip" where players threw balls at a metal bull's-eye in order to dunk a black youth seated above a tank of water.

charged, "There are more gambling devices per square foot at the State Fair of Texas than there are in Reno," but after three days of investigation, District Attorney Henry Wade permitted 32 games to reopen.

1952

With the Korean War stalemated and congressional investigations dragging on, Americans wearied of serious issues and settled back to partake of the good life — a frivolous '50s mix of 3-D movies, panty raids, poodle cuts and paint-by-number pictures.

In 1952, people discovered how much they liked Ike and loved Lucy. Fairgoers found still another object for their affection — the Big Texan, a four-story high, hook-nosed, wolfish grinning cowboy figure, dressed in Lee Riders and standing in the middle of Grand Avenue.

Notable newcomers to the entertainment lineup included a daily three-ring circus, with proceeds benefitting the Variety Club Boys Ranch; Dean Martin and Jerry Lewis in a vaudeville-style review; and pro football in the Cotton Bowl. From the remains of the Boston Yanks, local ownership put together a new NFL franchise called the Dallas Texans. The infant Texans, unfortunately, had trouble winning games, drawing fans and paying bills. They finished the season as a road team, moved to the east coast and enjoyed later success as the Baltimore Colts.

The centrally-located exhibit building with the 170′ tower was renamed the Electric Building in 1952. Women were attracted by the latest in freezers, home laundry equipment

In the free-wheeling years after the war, merchants in Kerens, Texas, had a problem. Residents of the tiny town were driving to nearby Corsicana or even 75 miles north to Dallas for pre-Christmas shopping sprees. Looking for a gimmick that might encourage people to spend money at local stores, the Kerens Chamber of Commerce built what they claimed was the world's largest Santa Claus, a 49-foot tall figure constructed from iron-pipe drill casing and papier mache with seven-foot lengths of unraveled rope for a beard.

The promotion was a big success during the 1949 holidays, but the novelty wore off the following year, and community support waned. In 1951, State Fair president R. L. Thornton purchased Santa's components for $750 and hired Dallas artist Jack Bridges to create a giant cowboy out of the material.

Big Tex made his debut at the 1952 State Fair of Texas. Wearing size 70 boots and a 75-gallon hat, Tex towered 52′ above wide-eyed visitors. His denim jeans and plaid shirt were donated by the H. D. Lee Company of Shawnee Mission, Kansas. Cosmetic surgery the following year straightened his nose, corrected a lascivious wink and allowed him to talk.

Big Tex's original location on the fairgrounds.

A large crowd gathered in front of the Hall of State to hear an address by Democratic presidential candidate Adlai Stevenson in 1952.

and garbage disposal units. Husbands tagged along to watch the World Series on dozens of TV sets including an enormous 27″-screen model.

Santa Gertrudis cattle, named for a creek running through the King Ranch, were showcased at the fair. The highlight of Mexico Day was a football game between the National University of Mexico and Austin College; Liberace headlined the East Texas Day Show in the Cotton Bowl; and Dr. Daniel A. Poling spoke at a religious festival in the stadium on the final Sunday.

Presidential candidate Dwight D. Eisenhower stopped briefly in Dallas without visiting the fair, but his opponent, Illinois governor Adlai Stevenson attended a private luncheon on the grounds and delivered a major address from the Hall of State Plaza.

For the tenth consecutive year, the State Fair of Texas broke its own attendance record.

1953

R. L. Thornton had completed eight terms as president of the State Fair of Texas and had no intention of giving the job up in order to run for mayor of Dallas. Only after obtaining a legal ruling that he could continue at the fair helm did Thornton accept a draft by the Citizens Charter Association, the political arm of the Citizens Council. He was elected to the city's highest office by a two-to-one margin.

The fair board approved the construction of dormitories in three livestock barns and voted to erect a Women's Building on the vacant lot where the Ford Building once stood. This project was slated to begin immediately after the 1953 exposition.

The directors also considered the plight of the Dallas Negro Chamber of Commerce. The organization was caught between its desire to continue working with the fair to sponsor Negro Day events and taunts of "Uncle Tom" from the black community. The fair and chamber finally agreed: "That the policy of the State Fair of Texas be altered to admit Negroes to amusement rides during the entire 16 days of the fair, on such rides where separate facilities are available and no contact is involved, and that this be accomplished by the posting of signs setting aside specific facilities for Negroes on said rides," — in effect, still separate, but somewhat more equal.

In 1953, a new program was developed to strengthen livestock shows at future fairs. Looking for a way to expand

In its first year, the Pan American Livestock Exposition attracted 200 guests from 12 Latin American countries. One of the highlights of the nine-day event was a Texas ranch party hosted by State Fair president R.L. Thornton.

markets for purebred animals and poultry raised in the southwest, the board turned to Latin American livestock producers and issued an invitation for a Pan American Livestock Exposition to be held during the first nine days of the fair. Nearly 200 guests from 12 Latin American countries attended this first-of-a-kind event and purchased $500,000 of livestock before returning home.

A coast-to-coast television audience joined 75,000 fans in the Cotton Bowl to watch Oklahoma top Texas on the opening day of the fair.

"More To See in '53" was a better slogan than "It's a LuLu in '52," and it called attention to a talented group of performers: Broadway's Queen of Musical Comedy, Ethel Merman, in a show tailored to her bombastic personality; singer Gordon MacRae with the Apache Belles from Tyler Junior College at the annual East Texas Day Show; and the military's largest all-woman musical organization, the United States WAF Band from Lackland Air Force Base in San Antonio.

Fairgoers were introduced to the new "mighty mite of electronics," a tiny transistor the size of a .22 calibre cartridge, which was hailed as the most important discovery in its field since the vacuum tube. A telephone without wires, called a "handie-talkie," illustrated one of the infinite ways transistors could be used. Innovative ironing machines were displayed as potential answers to that one household chore most resistant to automation.

Information about jet aircraft was available, and new farm implements, such as the subsoiler and the one- and two-row cotton picker, were displayed on the outdoor terraces. Dancing

Waters, an aquatic ballet, utilized 19 motors, 4,000 jets, 60,000 watts of power and 38 tons of water in presenting each show. But the real technological marvel of 1953 was one that even the youngest fairgoer could appreciate — Big Tex talked. The smiling cowboy flapped his giant jaw and rumbled a friendly "How-dee, Folks!"

In recognition of the growing importance of the fashion industry to Texas, runway shows were staged daily which honored 85 winning garments as selected by a panel of judges.

Kids attending Rural Youth Day gobbled down 90,000 hot dogs, or as someone calculated, 24 miles of hot dogs laid end-to-end. When the mustard ran low, a police escort was required to bring in a fresh supply. Total mustard consumption was 130 gallons.

A heavy rain on the final day ended the run of attendance records, but fair officials were quick to point out that 1953 ranked number two on the all-time list.

1954

Without delay, work began on the $500,000 Women's Building. Structural planning had been the responsibility of an all-male committee. Leah Jarrett, in charge of women's activities at the fair, said tactfully, "It was not the type of building women would have designed." For one thing, the low ceiling prevented vertical display of large hooked and braided rugs. And while air conditioning was certainly appreciated, the exposed machinery detracted from the building's appearance. These objections, far from representing female ignorance of construction matters, pointed up two glaring oversights that would frustrate designers of future trade shows.

The anticipated move of the women's department freed the Horticulture Building and surrounding area for development of a Garden Center. This project would be cooperatively financed by the fair, Park Board and the 120 Garden Clubs of Dallas.

Dallas Power and Light provided a welcomed addition to the park. The utility company installed a clock and signs that spelled out the word "ELECTRIC" on the tower above that exhibit building.

Fair Park was changing and so was Dallas. The city's population now approached the half million mark. Conrad Hilton was building a 19-story, modernistic blue-green colossus of a hotel downtown, and plans for a new $5 million terminal at Love Field were on the drawing board.

A series of decisions and discoveries in 1954 would touch the core of American life. The French lost the Battle of

Early in 1954, Dallas Power and Light Company donated a distinctive clock and sign to identify the building used to house its exhibits each fall.

Dienbienphu and were driven out of Indochina; Vietnam was divided into two countries. Dr. Jonas Salk developed a vaccine for polio, and the American Cancer Society reported test results that showed a higher cancer death rate for cigarette smokers. Finally, the United States Supreme Court ruled in the case of "Brown vs. Board of Education of Topeka" that segregation in public schools violated the 14th amendment.

Oscar Hammerstein II arrived in October for the State Fair engagement of "The King and I." His partner, Richard Rodgers, remained in Hollywood to supervise the filming of "Oklahoma." The fair show opened with a strong advance sale, and word of Yul Brynner's magical performance created an even greater demand, especially for the $4.80 orchestra seats.

Fairgoers lined up to see the fabulous $10 million Hope Diamond, a carefully-guarded focal display in the new Women's Building. Lines also formed outside the House Beautiful model home, built especially for the fair as a standard against which "people may measure their own homes, their own way of living and their own dreams for the future." The all-electric dwelling was designed by architectural students at the University of Texas and would be featured in an upcoming issue of *House Beautiful* magazine. One of the highlights was an undercounter dishwasher, an exciting improvement over free-standing models that had to be rolled up to the sink.

In its second year, the Pan American Livestock Exposition attracted 450 Latin Americans from 16 countries who made livestock purchases in excess of $2 million. Mayor Thornton, in his dual official roles, hosted a party for all the foreign guests at his own ranch outside Dallas.

Livestock activities during the second week of the fair centered on junior shows and auction sales. One first-time exhibitor, a 16-year-old girl, earned $510 at the junior turkey show for three broad-breasted bronze hens that had cost 76 cents each six months earlier.

Rows of temporary benches allowed the Texans and Oklahomans to erase the old stadium record, but it was the rural youth, arriving in 3,500 bright yellow school buses on October 16, that pushed single day attendance to a new high just under 300,000.

A national survey in 1954 indicated that the average American's favorite meal consisted of fruit cup, vegetable soup, steak and potatoes, peas and pie á la mode. Texas fairgoers that year abandoned worries about a well-balanced diet, and in 16 days consumed an estimated 1.5 million hamburgers and hot dogs, 140,000 boxes of fried chicken, 200,000 candied apples and 500,000 bags of popcorn. Food vendors in 300 locations served everything from 10 cent snow cones to $3.50 beef dinners.

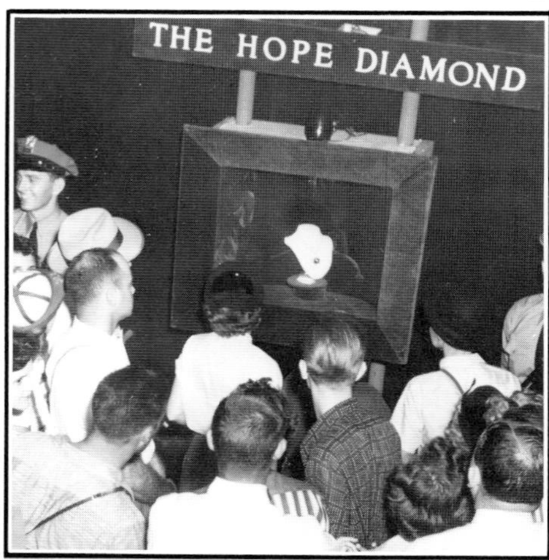

(top) Fairgoers crowded into the new Women's Building for a look at the $10 million Hope Diamond.

(bottom) Hopalong Cassidy of TV western fame made a Cotton Bowl appearance in 1954.

Daredevil shows and thrillcades were a staple of fairtime entertainment. These events were staged in front of the grandstand built in 1934 during the brief revival of legalized betting on horse races.

Attendance topped 2.5 million. The net profit was down slightly from the year before, but as Bob Thonton reminded, "The fair is not trying to see how much money it can make, but how much entertainment it can provide, and how many people it can attract to the City of Dallas." Privately, Thornton admitted that it was getting very difficult to keep breaking records.

1955

Everybody was singing about Texas and Texas heroes in 1955. "The Ballad of Davy Crockett" stayed on the charts for 20 weeks, and radio stations played Mitch Miller's up-tempo version of "The Yellow Rose of Texas" from August through November. The year's number one record, however, was "Rock Around the Clock" by Bill Haley and the Comets, a portent of things to come.

Except for normal maintenance and repair, the State Fair of Texas was willing to let the dust settle in the park during the off-season. There was preliminary talk about adding another 5,000 seats to the Cotton Bowl, and John Carpenter continued his pursuit of funding for the multi-million dollar Livestock Coliseum, but the fair directed most of its energies and resources toward upgrading the cultural and commercial programs. A comprehensive exhibit of color television was mounted, and Great Britain, France, Finland, Sweden, Belgium, Japan, India and West Germany were represented in a cluster of displays at one end of the General Exhibits Building.

A Rayburn-Johnson Day was held to honor the powerful Texas duo serving in Washington as Speaker of the House and Senate Majority Leader. The fair also paid tribute to Houston publisher and former Secretary of Health, Education and Welfare, Oveta Culp Hobby.

The Dallas Museum of Fine Arts played host to photographer Edward Steichen's "Family of Man" exhibition during the 1955 fair.

Free admission tickets were provided to a half million school children around the state, and teachers were encouraged to bring their classes to the fair on field trips.

One show of exceptional merit and educational value had been threatened by both censorship and cancellation. The Dallas Museum of Fine Art had scheduled Edward Steichen's acclaimed "Family of Man" during the fair period, but right wing interests demanded that works by Russian photographers be removed from the exhibition. Steichen was adamant that the collection be shown intact or not at all. The museum withstood the criticism and presented the show as Steichen insisted.

The Twister and the Scrambler introduced new thrills on the midway. Fairgoers packed the mile-long maze of rides, games and food stands throughout the first weekend. Another huge throng arrived on Music Festival Day. The crowd was a colorful sea of ducktail haircuts, letter jackets and band uniforms. Shortly before noon, a seat near the top of the Sky Wheels broke loose and spilled its three occupants through the framework of the ride to the operational ramp below, a fall of nearly 100′. A 15-year-old girl from Memphis, Texas, was pronounced dead on arrival at Baylor Hospital. Her companions suffered serious injuries. It was the first fairtime ride fatality in more than 60 years.

Later in the week, a group of students from Lincoln High School made an unsuccessful attempt to break down racial barriers on High School Day. The following Monday, Negro Achievement Day, 22 members of the NAACP Youth Council picketed outside the gates to protest midway ride policies. Their signs read: "Don't Trade Your Pride for a Segregated Ride" and "This is Aggrievement Day at the Fair — Stay Out!" They passed out material charging that Negroes were routinely barred from two rides and all restaurants, while being placed at the back of other rides. "If you visit the fair as we did on any other day," the pamphlet warned, "you will be humiliated and disgraced as we have been."

In response, President Thornton referred to the 1953 agreement that had opened all but two rides to Negroes. He explained that these rides were restricted because it was felt they might create a situation that would lead to physical violence.

"For the balance of the fair," Thornton announced, "these rides, previously restricted, will be open to all people. With reference to certain restaurants, we have moral and legal commitments which will prevent any change in policy . . . After the close of the 1955 fair, we will be happy to meet with representatives of the Negro Chamber of Commerce and re-examine the entire matter."

1956

The Supreme Court had ordered schools desegregated "with all deliberate speed," but southern congressmen were urging states to resist "by all lawful means." A bus boycott in Montgomery, Alabama, led by a young Baptist minister named Dr. Martin Luther King, Jr., had demonstrated the effectiveness of economic protest.

In a letter to R. L. Thornton written in the spring of 1956, the president of the Dallas Negro Chamber of Commerce,

Visitors were introduced to a futuristic transportation system in 1956, when a short monorail line was built inside the park.

M. M. McGaughey, explained that organizations's withdrawal of support from the fair: " . . . we understood that you and your directors had gone as far as you could in making the adjustments that we desired and which we believe you felt were reasonable in the light of present day procedures. We sincerely hope that you appreciated that we also had retained our position in the face of the most bitter criticism from our community and that loss of the support of our own community would have rendered our organization worthless . . ."

The directors decided to go ahead with plans for Negro Achievement Day and approved a $5,000 expenditure to secure the inimitable Louis Armstrong as the special entertainment feature.

At Thornton's recommendation, board members pared the budget of some costly "extras" including the annual religious festival, goodwill trips to Latin America, the Distinguished Texan Award and rain insurance. Gate admission was raised from 60 to 75 cents. Then the 76-year-old fair president offered his newest idea: America's first commercial monorail line to be installed and operating by opening day.

The system cost $125,000 to build, money loaned directly or guaranteed by the fair. From near the front of the park, the elevated line would extend along First Avenue to a loading station on Cotton Bowl Plaza. The fare to ride in the 60-passenger, fiber-glass coach was set at 25 cents.

The Monorail was described as "safe, silent and swift," but first day visitors to the fair could only watch workmen, sweltering in 96 degree temperatures, as they pressed to finish construction. By the end of the first week, test runs were made carrying loads of bricks, and eventually paying customers got a brief trip on the transportation system of the future. There was talk of building a monorail line to bridge the 39 miles between Dallas and Fort Worth.

Fairgoers were treated to other spectacular sights at the 1956 exposition. The fair and Dallas Power and Light had improved the Esplanade with dramatic illumination and multiple fountains, and Big Tex had acquired a pet — a 12′ tall plastic model of a Hereford steer. The animal, created by the Ralston Purina Feed Company, had an audible repertory of bellows and moos and a hollow interior so that people could watch his insides in action. Scaled displays demonstrated how feed is transformed into beef, how milk is produced and how a calf develops from embryo to birth — the latter processes, significant accomplishments for a steer of any size.

Those who wanted "name" entertainment found it. Broadway's Bobby Clark and TV western star Allen Case appeared in a touring company production of "Damn Yankees." Keyboard comedian Victor Borge performed in the

"The Champ," a hollow Hereford made of plastic, measured 12′ tall and weighed 4,000 pounds.

Elvis Presley in the Cotton Bowl at the 1956 State Fair.

Cotton Bowl. Louis Armstrong gave four concerts in one day without problems or protests. And a 21-year-old, former truck driver from Memphis, Tennessee, electrified a screaming, predominantly female audience of 26,500 with his orgiastic singing style. Elvis Presley, protected from the stadium crowd by a 10′ wire fence erected around the perimeter of the playing field, rocked his way through "Heartbreak Hotel," "Don't Be Cruel," and a string of hits-to-be.

Elvis' fans were not the only deliriously happy fairgoers. Oklahoma Sooner supporters went wild over a 45-0 stomping of Texas.

State Fair management was pleased with 1956's results and 1957's prospects, which included the smash musical hit "My Fair Lady." Charlie Meeker was promoted to the position of assistant general manager, and supermarket executive Bob Cullum joined the board of directors. Arthur Hale succeeded the late S. Bowen Cox as fair secretary. And to eliminate confusion with the new Memorial Auditorium, the dowager of Dallas theaters was renamed State Fair Music Hall.

1957

Dallas was ripped by a vicious tornado in the spring of 1957, and just before the fair opened, U.S. complacency was shaken by the launching of Sputnik and the use of federal troops to integrate Central High School in Little Rock.

The acquisition of eight blocks of real estate facing Pennsylvania Avenue increased the size of Fair Park by 13 acres and created a much-needed 2,000-vehicle auxiliary parking lot.

Elsie returned to the fair, this time with twins, and on the midway a 42″ midget billed himself as the world's smallest concert violinist. But despite such spectacles, fairgoers showed increased interest in serious exhibits. With the space race on, the Navy offered a first look at a scale model of the satellite it hoped to put in orbit, and the Army's display focused on the potential of its guided missiles, the Nike Ajax and Nike Hercules.

The women's department scored a coup by arranging for handmade rugs and coverlets, dating to colonial times, on loan from the Metropolitan Museum of Art in New York. A collection of early American silver was presented by Towle, and the Singer Sewing Maching Company took over sponsorship of the fashion show with popular local radio personality Elizabeth Peabody as commentator.

The Singer Sewing Machine Company sponsored daily fashion shows in the Women's Building.

(above) Two small paddleboats operated on the lagoon away from the noise and bright lights of the midway.

(left) Actor Ronald Reagan paid a visit to the fair in 1957 and presented awards to winners of the firemen's pumper races.

The Monorail, now fully operational, attracted passengers because of its novelty — as a ride, it was neither very exciting nor very long. Transportation from another era was represented with the introduction of two replicas of paddle-driven Mississippi showboats on the lagoon. Each of the small crafts was designed to carry 30 persons.

"My Fair Lady" played to an audience of 88,379 over its 24-performance run, doubling attendance at the previous year's show. The customary 75,000 inside the Cotton Bowl saw Oklahoma coach Bud Wilkinson give his former pupil, new Texas coach Darrell Royal, a 21-7 lesson. Ice Capades continued to draw well, as did the Thrillcade in front of the grandstand. Spectators at the opening show of this motorized mania watched in horror when an engine fire turned one of the drivers into a human torch. The unlucky stuntman was hospitalized with second and third degree burns.

The 1957 exposition was plagued with an old nemesis — stormy weather. The clouds opened up and soaked the park on five separate occasions including the traditionally-big middle Sunday. The Pat Boone Show, scheduled for the Cotton Bowl, had to be moved into the covered Livestock Pavilion.

The rain provided a respite from attendance records, but the fair maintained its solid financial base. The organization was current with all obligations and comfortably ahead on bond payments. The city showed signs of being ready to build the long-delayed Coliseum, and Bob Thornton was talking about another of his dreams — a massive World Trade Building to be constructed in Fair Park, probably on the Band Shell or swimming pool site, and operated as a year-round asset for Dallas business and a public attraction at fairtime.

1958

A Fort Worth wrecking contractor bid $2,500 for the salvage value, and the grandstand came tumbling down in the spring of 1958. Built in the 1930s as part of the racetrack complex, the aging structure followed the Globe Theater and original Fine Arts Building into Fair Park history, razed to meet demands for more parking and newer facilities. The picnic pavilion and swimming pool would be next.

Construction of the Coliseum got underway at last. In the final set of blueprints, the architects designed a vast, modern building appropriate for livestock shows, circuses, rodeos and other large-scale events requiring a packed dirt surface. Seating capacity adjusted from 7,000 to 11,000 by adding chairs on the arena floor. The structure was heated, though not air conditioned. Turquoise-colored metal panels covered the exterior in bold contrast to the traditional, southwestern-styled buildings nearby. The $1.9 million project, financed by the City of Dallas, also included an adjacent stock barn and exercise ring. Completion was expected in time for the 1959 fair, and the first major off-season booking would be the National Finals Rodeo during Christmas week of that year.

In 1959, after years of delay, the City of Dallas built a $1.9 million Coliseum on the site of the old racetrack and grandstand.

Nationally in 1958, kids caught the Hula Hoop craze, Congress approved Alaskan statehood, scandal swept television quiz programs off the air, and women challenged male-only admission policies at Texas A&M.

Locally, the State Fair Musicals had a disastrous summer season, finishing $75,000 in the red, and R. L. Thornton, in his third term as mayor and 14th year as fair president, warned that losses of this size could not be tolerated.

Thornton was championing another cause — "Grand Ol' Texas," a 6.5 acre, year-round amusement attraction to be built in Fair Park. Conceived by theatrical designer Peter Wolf, Grand Ol' Texas incorporated a half-scale Alamo with a smoky battle raging inside, the cave of train robber Sam Bass with "bandits" taking mock shots at visitors, a steamship modeled after the Trinity's *S. S. Harvey,* a "boom town" from Spindletop days, a frontier saloon and rows of quaint shops. Wolf had announced the $4-5 million project earlier, but had been unable to secure financing, and Thornton now proposed that the State Fair become a partner in the venture and that the city sell revenue bonds to pay for construction. A tentative agreement was reached that the city would issue the bonds when the fair could show $150,000 worth of signed leases from concessionaires.

The 1958 State Fair of Texas featured a "Shower of Stars" with Tennessee Ernie Ford, George Gobel, Red Foley, Tito Guizar and "Sky King" and "Penny" from the popular TV

"Grand Ol' Texas," an imaginative western-themed amusement park, proposed and planned, but never built in Fair Park in the late 1950s.

The Texas International Trade Fair, organized in 1958, opened the door for Dallas' emergence as a market center for foreign products. Early participants included: (top left) Great Britain, (top right) Guatemala, (center left) Egypt, and (center right) India.

(right) The years following World War II were marked by unparalleled growth for the State Fair of Texas. Attendance rose from a prewar high of 1,252,527 (1941) to a record 2,801,305 in 1959.

series. Composer Meredith Willson accepted the first Texas Music Festival Award, and "The Music Man" was a blockbuster favorite with fairtime audiences.

Visitors in a buying mood found quality foreign products at the first Texas International Trade Fair in the General Exhibits Building. Those wanting to learn about rural life in

Texas studied a giant display: "Agricade — A Generation of Farm Progress," and anyone looking for a little Lone Star culture gathered in front of the Hall of State on weekends for concerts by the Texas Boys Choir.

Two changes were noted and approved by fairgoers. The Electric Building was now air conditioned, and Big Tex had undergone cosmetic surgery. The tall fellow's chin and nose had been sculpted to more aesthetic proportions.

Attendance passed the 2¾ million mark, but Thornton, ever aware of the balance sheet, noted that more people had not translated into more profit. He recommended raising admission from 75 cents to one dollar and also suggested to the board that it might be time to start grooming his successor.

1959

The fair invested $54,000 in plans and promotional efforts for Grand Ol' Texas, but the theme park within-a-park ran into what novelist Joseph Heller later labeled "Catch 22." The city was willing to provide financing as soon as enough concession leases were signed; the prospective leasees wouldn't okay contracts until the city committed the money. A point of no progress existed, and the fair finally decided to take its losses and withdraw from the partnership.

Work progressed on the Coliseum with one sad note. John W. Carpenter, who had given 15 years to the project, died in June. The building was finished in September.

Trammell Crow opened the Dallas Trade Mart, the second phase of his market complex, in 1959, and Lamar Hunt, frustrated in his efforts to land a National Football League franchise for Dallas, organized the American Football League.

In July, Vice President Richard Nixon engaged in a "kitchen debate" with Soviet Premier Nikita Khrushchev while visiting a trade exhibition in Moscow. Three months later, the vice president rode in the opening day parade for the 1959 State Fair of Texas. With his wife, Pat, Nixon cut the ceremonial ribbon, shook hundreds of hands and visited all 18 booths of the second Texas International Trade Fair.

Exposition visitors were invited to see themselves on color television as one of the highlights of an expanded exhibit program. The City of Nikko featured tiny replicas of 24 Japanese shrines and pagodas. Merrill Lynch sponsored a miniature stock exchange with ticker tapes and up-to-the-minute quotations. And in the Varied Industries Building, another Fair Park structure with an updated name, a wide assortment of money — from old "saddlebag" currency to an

Vice President Richard M. Nixon and his wife, Pat, cut the ribbon to open the 1959 fair.

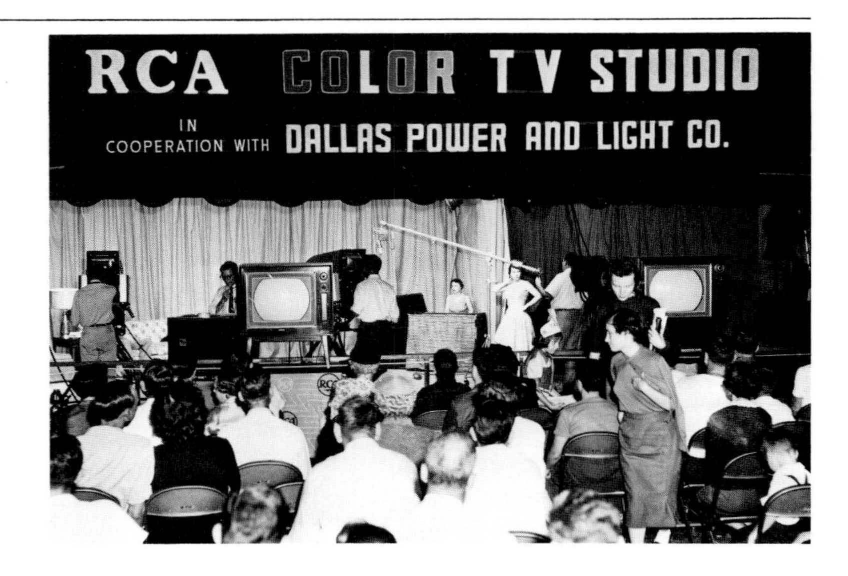

Many Texans saw color television for the first time while visiting the fair in the late 1950s.

actual $100,000 bill — was displayed under glass.

The McGuire Sisters had top billing for "Star Light! Star Bright!" a musical review which also showcased the bright young comedy team of Rowan and Martin. Free outdoor shows during the fair featured the talents of Herb Shriner, mambo king Perez Prado, television actor Robert Culp, Woody Herman, Sam Cooke and Mitch Miller.

With large crowds, good weather, no protests and no accidents, the 1959 edition appeared to be one of those occasional fairs where everything went exactly as planned — until area hospitals admitted 40 persons with symptoms of food poisoning, and all 40 reported having visited the fair. The victims recovered without complications. An investigation revealed that most of them had eaten chocolate eclairs at the Sidewalk Cafe in the General Exhibits Building. A batch of cream filling prepared by an employee with an infected finger was cited as the probable cause, and health authorities banned further sale of cream-filled pastries.

The final accounting showed 2.8 million fairgoers and a healthy $400,000-plus in net profit. The dollar totals were especially impressive in light of the losses taken on Grand Ol' Texas and the $60,000 deficit in the summer Music Hall operation. The new Coliseum functioned flawlessly for fairtime horse shows, but the first attempt to use the facility for a major revenue event disappointed everyone involved. The National Finals Rodeo attracted followers of the sport from all across the country, but Dallas residents showed little interest and made other plans for the Christmas season.

The State Fair Board acceded to Thornton's request and selected one of its own, C. A. Tatum, as a trainee for Uncle Bob's job. The advent of the new year and new decade marked another milestone in State Fair history. It had been 75 years since Billy Gaston, Sydney Smith, Alex Sanger and the others built the foundation for a multi-million dollar empire on 80 acres of hog wallow prairie.

Silver Threads Among the Gold

1960-1972

"The times they are a-changin'," sang Bob Dylan, and John Kennedy talked about a New Frontier. According to the 1960 census, the population of the United States had reached 180 million. The economic strength of the country was shifting westward. Businesses allied with the vital aircraft and electronics industries were establishing a power base in North Texas. A six-lane turnpike now linked Dallas and Fort Worth. Commercial jetliners flew out of Love Field, and the area's first covered shopping mall, Big Town, blended convenience with comfort. Rush hour congestion on Central Expressway was accepted as an ambiguous sign of progress.

1960

In the spring of 1960, the Park Board announced that the old swimming pool in Fair Park would be demolished to provide additional parking for the Musicals' summer season. According to records, fewer than 20,000 had used the 35-year-old facility in 1959, most swimmers preferring one of the city's 13 newer pools. "The area around Fair Park has become industrial instead of remaining a residential section," explained one official.

The area, in fact, remained largely residential, but predominently black. With lunch counter sit-ins and other forms of nonviolent protest gaining strength across the south, many observers felt that the pool, though arguably obsolete and deteriorating, was razed at this time to avoid confrontation.

South Dallas, once populated by prosperous families who built mansions on its tree-lined streets, had gone through cycles of change while Fair Park developed along its eastern boundary. After the wealthiest residents moved on, the

Fair Park's swimming pool, built in 1926, was torn down in the spring of 1960.

The traditional downtown parade kicked off opening day activities at the Diamond Jubilee Exposition celebrating the fair's 75th anniversary.

neighborhood became the center of social and religious life for the city's large Jewish community. Following World War II, many of these homeowners relocated to North Dallas, and black families began moving into the area. When Forest Avenue High School closed in 1956, and the building reopened as James Madison, a Negro school in the city system, the passage from a white to black neighborhood was virtually complete.

While parking on the fairgrounds was always an important consideration, there was some question about any immediate need for additional spaces near the Music Hall, since attendance at the summer shows had not improved. The State Fair Musicals lost $116,000 in 1960, and at the end of the season, managing director Charlie Meeker accepted a position with the yet-to-be-built Cary Plaza Hotel.

The Cotton Bowl acquired not one, but two new tenants that fall. Both the Dallas Texans of the fledgling American Football League and the fledgling Dallas Cowboys of the established National Football League played home games in the stadium. A schedule was drawn up apportioning dates between the two teams, although their combined crowds could easily have fitted into the 75,000-seat Bowl on any given Sunday.

The State Fair of Texas announced a Diamond Jubilee Exposition, but the 75th anniversary edition did not differ greatly in content from the fairs immediately preceding it. Popular features were repeated, new stars fitted into familiar formats and a few names changed to reflect the nature of the celebration.

"Flower Drum Song" kept the Music Hall box office busy, and composer Richard Rodgers received the annual Texas Music Festival Award. Other entertainment headliners included Fabian, Brenda Lee, Homer & Jethro, Emmett Kelly and Jimmie Rodgers.

Senior citizens were admitted free on Jubilee Day. Nelson Eddy sang medlies of sentimental favorites, and management paid tribute to persons who had attended the exposition during its first five years, 1886-1890. A group of 60 were feted at a luncheon.

Arthur Godfrey and his Palomino, Goldie, appeared at shows for hunters, jumpers, gaited, fine harness, parade, road and walking horses. Cowboy star Rex Allen and Ko-Ko visited the next series which featured quarter horses, Palominos, Arabians, Shetland ponies and an open cutting horse competition.

The Texas-OU game had outgrown its limited definition as a football contest. Nor could night-before activities in downtown Dallas accurately be called pep rallies. Since 1958, a dance had been sponsored in Memorial Auditorium to keep the celebrating under control, but this year, when the party

Joan Robinson Hill took part in the State Fair Horse Show in 1960. The Houston socialite's death nine years later was the focus of a sensational murder trial and inspired the national bestseller "Blood and Money."

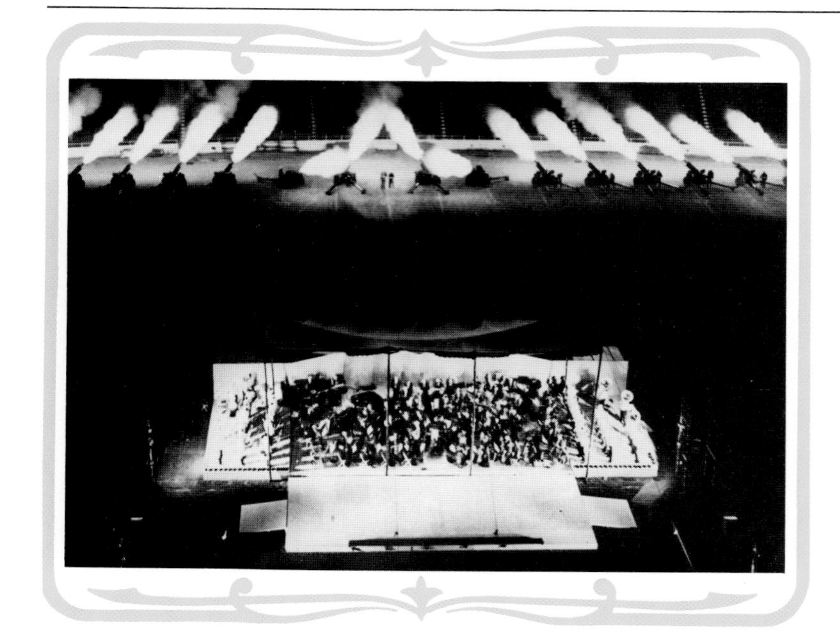

A tradition was born when the Dallas Symphony Orchestra concluded an evening concert in the Cotton Bowl with Tchaikovsky's stirring "1812 Overture" dramatized by cannonading on the field and the burning of Moscow in fireworks.

broke up at 2 a.m., thousands poured into the streets creating a scene of Mardi Gras proportion and revelry.

The Cotton Bowl was busy nearly every night. In addition to four major college and professional games, the fair scheduled Gil Gray's International Circus, with a full complement of lions, elephants, bears, jugglers and wire-walkers, followed the next evening by the first Dallas Symphony Spectacular. The orchestra, supplemented by the Air Force Band of the West from San Antonio, offered a thunderous rendition of Tchaikovsky's "1812 Overture." Special effects at the climax featured real artillery fire from a battery of Marine howitzers and a barrage of fireworks.

The fair advertised the second Monday simply as "Achievement Day," but the special program for Negroes remained essentially unchanged, and pickets again marched outside the gates. General manager Jimmie Stewart defended his organization, saying it offered Negroes more privileges and equality than any other institution in Texas. This may well have been true. Given the times, the State Fair had shown significant concern and respect for black interests, but as a highly visible event, the fair was a natural target for protest against the discriminatory policies it perpetuated and those practiced in the community as a whole.

It was not an easy fair for management. Two young men were wounded in a shooting on the grounds. A snakehandler on the midway was bitten by a poisonous moccasin. And inclement weather kept attendance below the accustomed record levels.

R. L. Thornton announced that he was "not retiring, resigning or surrendering," but after eight years as Mayor of Dallas, he declined to run for office again in the spring. He was not ready, however, to relinquish the fair presidency.

1961

In his inaugural address on January 20, 1961, President John F. Kennedy spoke of challenges and declared that a new generation of Americans was ready to accept responsibility.

The State Fair of Texas was entering a transitional period. There was no question that the 80-year-old Thornton was still running the show. He remained an overwhelming personality and master of detail, to the extent that he insisted on a daily in-person report from general manager Stewart. But the fair, though still something of a "Mom and Pop" operation with regard to record keeping and accounting procedures, had become a big business that required an adequate substructure of salaried managers. Gradually, the State Fair staff was given more authority to plan and conduct business, and job openings were filled by promotions from within the organization. Tom Hughes, barely 30, but with experience that dated back to selling cushions at the first Starlight Operettas, was appointed managing director of the Musicals. Joe Rucker was named assistant to the general manager. Rucker, in his nine years with the fair, had pioneered programs in foreign exhibits, special events and grounds decoration.

In the spring, one week before the Dallas Independent School District announced it would comply with court-ordered integration, R. L. Thornton told his board of directors that the fair would obey "the law of the land" and do likewise, beginning with the summer midway season.

Freedom rides and demonstrations made headlines in 1961. Newspapers also reported the Bay of Pigs debacle, the Berlin Wall, America's first astronaut in space and Roger Maris' assault on Babe Ruth's homerun record. But a growing number of Americans were turning away from printed sources of information and following events on television. Over 55 million

For some, a chance to watch the World Series was the biggest attraction in the appliance-filled Electric Building.

sets were in operation, and a Canadian scholar named Marshall McLuhan was writing, ironically, a book to explain the media revolution.

Television, a "vast wasteland" and general whipping boy in 1961, got part of the blame for another bad summer at the State Fair Musicals. Thornton had too many plans for the fair's future to permit this continual drain on its financial resources. Of course, not all of the redoubtable president's ideas were inspired. When he recommended that Big Tex be rebuilt as a permanent statue, board members questioned everything from the cost to whether the giant cowboy's "look" was really appropriate for a monument.

Big Tex put in an early appearance that year. Workmen set him up in September for his role in the movie "State Fair," but tailing winds from Hurricane Carla tore his clothes, and the production crew was forced to shoot around him until repairs could be made.

"Exposition of Music" was the official theme of the 1961 State Fair. The concept was carried out by contests for old fiddlers, a Gospel Song Fest, Sunday jazz concerts, a western music jamboree, the statewide band festival, a second "1812 Overture" performance by the Dallas Symphony, and Rodgers and Hammerstein's "The Sound of Music," only two years removed from its Broadway opening, starring Florence Henderson as Maria.

Fallout shelters were on display, and Zale's Jewelry Stores exhibited a replica of the United States Capitol made of 217,569 cultured pearls. For the most part, however, regular fairgoers saw what they had seen before. Certain events were predictable: kids from farming communities consumed stupendous quantities of free hot dogs; 150 celebrants were arrested for drunkenness before the Texas-OU game; someone complained that the cold-eyed strippers in the girlie shows were corrupting Texas youth; and thousands flocked to the Automobile Building where new cars were more alluring and more expensive than the year before. Prices had climbed into the five-figure range for the most luxurious models.

But in the eyes of at least one visitor, everything at the fair was new and wonderful. Bashir Ahmad, a camel cart driver from Karachi, attended the exposition as the guest of Vice President Lyndon Johnson. With Johnson as his guide, Bashir learned about Texas heroes at the Hall of State, watched a tractor competition in the Coliseum and spent time talking to youngsters who were preparing their sheep and cattle for show. The visit, an outgrowth of a chance meeting during the vice president's trip to Pakistan six months earlier, concluded with the tall Texan presenting a new pickup truck to the tiny cart driver.

"When good fortune catches hold of a person, good fortune takes him all the way," said Bashir.

Vice President Lyndon B. Johnson invited a Pakistani camel cart driver to attend the 1961 State Fair.

Hollywood's first, black-and-white, straight dramatic version of "State Fair," adapted from Philip Strong's 1932 novel, starred Janet Gaynor, Lew Ayres, Will Rogers and Louise Dresser. The next "State Fair," a 1945 technicolor release with Jeanne Crain, Dick Haymes and Dana Andrews, boasted an original musical score by Rodgers and Hammerstein.

But the "State Fair" that Texans remember most was filmed in Fair Park one month prior to the opening of the 1961 State Fair of Texas. The cast featured recording stars Pat Boone and Bobby Darin, former *Vogue* model Pamela Tiffin, youthful songstress Ann-Margret, screen veterans Tom Ewell and Alice Faye, plus an 800-pound California hog in the role of "Blue Boy," a sow from a nearby Plano farm portraying "Zsa Zsa" and Big Tex as himself.

The 20th Century-Fox film was directed by actor Jose Ferrer and produced by Charles Brackett whose screen credits included "Lost Weekend" and "Sunset Boulevard." Brackett had a $4.5 million budget. Additional location shooting was scheduled for two farms near Kemp in Kaufman County, the fairgrounds racetrack in Oklahoma City and a trailer camp in California's San Fernando Valley, but most of the action took place in Fair Park from the flower-bedecked Esplanade to the ignoble Swine Pavilion.

Pop singer Darin brought his very-pregnant wife, actress Sandra Dee, to Dallas. Boone, the former North Texas State student known for his white bucks and squeaky-clean image, was accompanied by his wife and four daughters. All of the stars participated in a special Music Hall benefit to raise funds for survivors of Hurricane Carla. The storm hit the Texas coast the same day that production began on "State Fair."

Bobby Darin managed another good deed while he was in town. Local disc jockey Ralph Chapman, who played the Harrigan half of KLIF's enormously popular "Murphy & Harrigan" team, wanted time off for a honeymoon, so Darin subbed for him on the early show.

Texans rushed to see "State Fair" when it was released, but the movie was outclassed and overshadowed by such excellent 1962 films as "Lawrence of Arabia," "The Music Man" and "To Kill a Mockingbird."

In the years that followed: "State Fair" settled in as a staple of late night TV; Bobby Darin died following open-heart surgery at age 37; Ann-Margret went on to superstardom; Pat Boone's daughter married Jose Ferrer's son; and deejay Chapman, using another first name, became the undisputed king of morning-drive radio in Dallas.

1962

More than 20 years had passed since the last major American exposition, and the public readily welcomed Seattle's Century 21 in the spring of 1962. That exposition's showpiece was a 600′ tower topped with an observation deck and revolving restaurant.

A group of State Fair officials journeyed to the northwest for a close look at this architectural wonder, and R. L. Thornton decided that Dallas needed a space needle of its own, which he naturally wanted to build in Fair Park.

Looking toward future needs, the fair's directors commissioned a study on midway renovation and expansion, but the projected cost was a jolting $12 million, and the idea was dropped for the moment. A series of architects and consultants would be employed to analyze Fair Park throughout the 1960s.

In response to an eloquent plea by Tom Hughes, the State Fair Board had granted the Musicals an eleventh hour reprieve and given Hughes two months to find underwriters for the summer season. Backed by commitments from businesses and individuals, the Musicals presented a solid package of shows, after which, with the fair's blessing and continued assistance in box office and bookkeeping operations, a separate and independent company was formed to be known as the Dallas Summer Musicals.

Formal themes had replaced rhyming slogans in pre-fair promotion, and the 1962 event was styled as "Exposition of Nations." This prompted an impressive new name for the oldest structure on the grounds: the World Exhibits Building. An international bazaar featured a broad selection of imported goods including items from Nationalist China.

Fairgoers, always with an eye for something new, found the exposition's Space Kitchen. An experimental feeding console illustrated how food and beverages would be provided for three space travelers on a 14-day voyage.

Fairgoers marveled at the speed and convenience of direct distance dialing.

Each year, Southwestern Bell sponsored one of the fair's largest and most popular exhibits.

Southwestern Bell showcased a replica of the new Telestar communications satellite. The telephone company consistently created one of the fair's most popular exhibits by combining demonstrations of direct distance dialing, a prize game called "phono" and the seductive offer to "hear your own voice."

Memorabilia from the glory days of the big top was spotlighted at Circusland. Photos and relics from the Hertzberg Collection of the San Antonio Public Library and the Circus World Museum of Baraboo, Wisconsin, were displayed along with a menagerie of animals under an old fashioned striped tent.

A 10-year-old girl from Marlin took both grand champion and reserve champion honors over 120 entrants in the junior turkey show, and a rooster named "Chicken," hatched under a lamp on a first grade teacher's desk, won a fourth place ribbon. Judges permitted the classroom pet, parentage unknown, to enter as a barnyard breed.

On the second Tuesday evening of the fair, 30 minutes before the East Texas Day Show was scheduled to begin in the Cotton Bowl, one of the small paddle boats on the lagoon exploded in flames. Screaming passengers leaped and dove into the shallow water. Fair officials and bystanders waded in to help them. The fire critically injured one woman and hospitalized 11 burn victims. Leaking gasoline was suspected as the cause, and the scenic water ride was closed permanently.

1963

In January, a thinner, white-haired R. L. Thornton, wearing thicker glasses, began his 19th year as president of the State Fair of Texas by assembling the board of directors to listen to a well-organized pitch for the $3.5 million space needle he wanted in Fair Park. The case was argued by Centennial architect George Dahl; Dr. Fritz Leonardt, a noted German engineer; John Bartel, who laid out cost projections; and Angus Wynne, who said he would construct it at his new Six Flags amusement park in Arlington, were it not for building code restrictions.

"It challenges me," said Thornton. " . . . it's a thrill people want. It's the country boy who wants to climb up on the tallest barn — the tallest tree." But this time the directors could not be swayed, and the space needle idea was tabled.

John, Paul, George and Ringo burst onto the pop music scene in 1963. A military coup overthrew the government of Ngo Dinh Diem in South Vietnam. Civil rights demonstrations increased across the country, and Martin Luther King,

Jr. led a march on Washington, where he told a crowd of 100,000 massed in front of the Lincoln Memorial: "I have a dream."

By late summer, his health failing, Uncle Bob Thornton announced that this would be his final year as fair president, though he would agree to becoming chairman of the board if such a position were created. Bob Cullum, president of the Dallas Chamber of Commerce, described Thornton as "a man who retires very slowly."

The 1963 State Fair was themed "Exposition of Our American Heritage." Flags for each of the 50 states flew from poles that bordered the Esplanade.

Fairgoers got a look at the new Southwestern Wax Museum, an entertaining collection of life-like figures, and the Age of Steam, an exhibit of locomotives and rolling stock from the golden era of railroads. Dr Pepper and Frito-Lay sponsored a daily free circus, and Lone Star Gas spotlighted self-cleaning, outdoor barbeque grills which featured controlled natural gas heat and permanent ceramic briquettes.

In keeping with the patriotic theme, an exhibit at the DAR House traced the development of the American flag. The Dallas Symphony and Dallas Civic Ballet combined for a tribute to John Philip Sousa in the Cotton Bowl. "Faces of Freedom," 75 paintings of historical events, was presented by John Hancock Life Insurance, and the colonial-costumed Deep River Fife and Drum Corps made a first-time fair appearance. Even the Music Hall show, "How To Succeed in Business Without Really Trying," was rooted in American tradition.

The middle weekend was devoted to football. The Dallas Texans had moved to greener pastures in Kansas City, but a three-day gridiron marathon saw SMU beat a Navy team

(above) The colorful Fife and Drum Corps from Deep River, Connecticut, performed at 1963's "Exposition of Our American Heritage."

(left) Clowns, acrobats and animal acts provided daily entertainment under an old-fashioned circus tent.

quarterbacked by Roger Staubach; number two in the nation, Texas, upset number one ranked Oklahoma; and the Dallas Cowboys squeeze out a 17-14 win over Detroit. The Friday night score from downtown was 168 jailed, 11 hospitalized.

For 16 consecutive days, it never rained a drop, and the final attendance count of 2,906,446 broke the 1959 record. Bob Thornton, who had missed most of the fair because of illness, showed up on the last night to congratulate the staff.

Four days after the fair closed, United Nations ambassador Adlai Stevenson, in Dallas for a speech, encountered a heckling, jeering, spitting crowd. One irate woman hit Stevenson in the head with the protest placard she carried.

And one month later, on November 22, 1963, as the presidential motorcade passed through downtown Dallas and turned on Elm toward the Triple Underpass, John Fitzgerald Kennedy was assassinated.

1964

The days that followed unleashed stinging criticism and contempt. The world indicted a proud city. Its leaders were stunned.

"Mr. Dallas," Robert L. Thornton, Sr., nearing the end of his life, was profoundly affected. In December, the State Fair's elder statesman retired, and C. A. Tatum became the organization's 24th president. Two months later, Thornton was dead.

C.A. Tatum, State Fair President, 1964-1965.

"Vision was his forte," eulogized the *Dallas Times Herald*. "No mountain was too high for Bob Thornton. He reached for the peaks and seldom missed . . ."

Felix de Weldon, acclaimed for his Iwo Jima statue in the nation's capital, was selected to create a memorial to Thornton. But a 250-acre monument already existed: virtually every structure on the fairgrounds had been constructed for the Centennial, an event Thornton secured for the city, or during his postwar leadership of the State Fair. And if the wily, homespun banker could have managed it, the equivalents of the World Trade Center, Reunion Tower and Six Flags Over Texas would have been built in Fair Park years before they became landmarks somewhere else.

In February, Jimmie Stewart announced that the Monorail would be dismantled to make way for a new form of aerial transportation, a 2,800′ Swiss Skyride. The tramway, with its 62 cage-like gondolas, would be comparable to the popular rides at Disneyland and the 1964 New York World's Fair.

Thousands of twinkling and colored lights, masses of chrysanthemums, new splashing fountains, lighted Tivoli arches, hanging fern baskets and towering maypoles characterized "Exposition of Lights and Flowers," the 1964 State Fair of Texas.

Governors from 16 southern states took part in the opening ceremonies and were welcomed to the city by Dallas mayor Erik Jonsson. Texas governor John Connally, fully recovered from his wounds incurred during President Kennedy's assassination, cut the ribbon, and Meredith Willson led the Longhorn Band in playing "Stars and Stripes Forever." Willson's latest show, "Here's Love," was the Music Hall attraction.

Fairgoers sampled their first Belgian Waffles in 1964. The crisp battercakes, topped with strawberries, whipped cream and powdered sugar, had been the walk-away snack hit at both the Seattle and New York fairs.

The Cafe de Paris, located outside the World Exhibits Building overlooking the Esplanade, provided another dimension in fairtime food service. In French provincial tradition, one hot dish, prepared by the chef at Dallas' elegant Old Warsaw Restaurant, was featured each day, with an assortment of French wines and cheeses.

The Natural History Museum presented a special exhibition, "Game Hunting in Texas," and WFAA-TV set up a studio in the Gas Building where viewers could watch live telecasts starring local favorites Julie Benell, Norvell Slater and Mr. Peppermint. A first-time feature in the women's department was called "Let's Face It" and offered demonstrations of cosmetics and hairdressing. National shows for Brahman breeders and Arabian horse fanciers headed the list of livestock activities.

But new or prestigious attractions were not enough to keep the State Fair on the front pages during the month of October 1964. Three Russian cosmonauts were orbiting the earth, while another Soviet, Premier Nikita Khrushchev, was toppled from power in a startling Kremlin shake-up. Dr. Martin Luther King, Jr. won the Nobel Peace Prize; President Johnson and Barry Goldwater fired last-minue pre-election salvos; and leaders from Dallas and Fort Worth met to settle their differences and begin serious discussions about a regional airport.

Workmen erect support towers for the new Swiss Skyride which replaced the Monorail in 1964.

Elegantly costumed riders brought a touch of glamour to the Coliseum during the Arabian horse shows.

1965

Lyndon Baines Johnson crushingly defeated his conservative Republican opponent in the presidential election. In the first months after Kennedy's death, this volcanic Texan had succeeded in pushing needed legislation and reforms through Congress; but 1965 brought increased American involvement in Vietnam, and despite passage of a Civil Rights Act, brutal resistance in Selma, Alabama, and a series of urban riots — most notably in the Los Angeles ghetto of Watts.

A magnificent, ultramodern shopping mall, North Park, opened alongside Central Expressway in Dallas that summer, and Sanger-Harris bucked a trend by building the first new downtown department store in 30 years. To the south, Houston unveiled the contemporary world's eighth wonder — the Astrodome.

In October, the State Fair saluted neighboring nations with an "Exposition of the Americas." The Royal Canadian Mounted Police presented their spectacular musical ride during the horse shows, and Danzas y Cantos de Mexico performed in the Cotton Bowl. In conjunction with the appearance of this renowned dance troupe, Big Tex wore a 15′ x 60′ serape, a gift from the Mexican government. Fifty craftsmen had labored to created this vibrant-colored blanket that weighed nearly 300 pounds.

The annual Texas-OU rowdiness escalated this year, and 371 of the downtown merry-makers spent at least part of the night in jail.

More ill-advised mischief ended tragically when three teenagers decided to take a fully-clothed swim in the Fair Park lagoon late on a Saturday night. The trio scuffled in the water about 30′ from shore, and police were able to rescue only two. An 18-year-old Dallas boy drowned in the incident.

Through the 16-day run, fairgoers marveled at the gigantic Budweiser Clydesdales, applauded magician Mark Wilson

The Royal Canadian Mounted Police Musical Ride — 1965.

and enjoyed fried chicken dinners at Youngblood's Restaurant in the newly-remodeled Old Mill Building. "Funny Girl," starring Marilyn Michaels as Fanny Brice, entertained Music Hall audiences.

Attendance rebounded after the slight dip recorded the previous year. Jimmie Stewart previously had announced his intention to retire after 15 years as general manager, and following the fair, C. A. Tatum revealed that poor health would force him to step down also. The organization selected a new leadership team. Joseph B. Rucker, who had served a long and thorough apprenticeship, was appointed general manager, and Robert B. Cullum, the dynamic and personable head of Tom Thumb Stores, was elected president.

In response to pressure from the Dallas Cowboys and, to a lesser degree, the colleges, the State Fair Board borrowed $350,000 to pay off the Cotton Bowl bondholders, thereby releasing the seat options this group controlled. An economic feasibility study of park redevelopment plans was initiated, and the directors approved an expenditure of $135,000 to convert the Coliseum into a multi-purpose facility with ice rink capability. Lamar Hunt was said to be working on a deal to transfer the New York Ranger hockey franchise to Dallas.

Joseph B. Rucker, Jr., General Manager, 1966-1972.

1966

In 1966, the first steps were taken toward formulating "Goals for Dallas," a concensus statement of the city's dreams and requirements. A segment of R. L. Thornton Freeway opened just north of Fair Park, and El

An artist's concept of Fair Park's future with waterways leading from the present lagoon to a pool in front of the Music Hall.

Bob Cullum and his younger brother Charles built the family's wholesale grocery business into a thriving supermarket chain. Considered one of the powerbrokers in city politics, Cullum lived in Highland Park, thus making him ineligible for elective office in Dallas. This arrangement suited the genial grocer, who said, with impish and exaggerated modesty, that he was meant to stay on the sidelines while others played the game.

Cullum was a State Fair aficionado cut from the same cloth as R. L. Thornton. There were ties between the two families. Bob Cullum and Bob Thornton, Jr., were friends and fraternity brothers at SMU, and when young Cullum needed money to get started in business, he turned to Thornton, Sr., for encouragement and a $250,000 loan.

Centro, the first of the Dallas County Community Colleges, offered classes in the building that once housed Alex Sanger's department store.

The fair settled on "Exposition of Young America" for a theme and tried to redesign its advertising campaign to reflect a youth-oriented approach. The result was predictably ludicrous. After years of schmaltz and corn, ads now featured pop art, allusions to Carnaby Street, a few hip phrases and a handful of mutant adjectives: gigantific, splendiferous, gaudful, funtastic and that old standby, dydamic. Beneath this strange promotional attire, the 81-year-old fair showed its customary colors.

Opening day coincided with the Texas-Oklahoma game, and night-before celebrating got completely out of control. Police reported 310 arrests, 110 injuries, 44 false fire alarms and 15 minor traffic accidents.

Big Tex wore a new LBJ-style hat to welcome fairgoers. The Varied Industries Building served as a theme pavilion spotlighting guitars, motorbikes and mod fashions. One booth sold T-shirts sporting bits of advice for the young: "Flunk Out Now, Avoid the Rush" and "Love Thy Neighbor, But Don't Get Caught."

Honors for the year's timeliest exhibit went to the U.S. Army which constructed a life-size replica of a Vietnamese village near the front gate. The display featured thatched huts, a simulated rice paddy, booby traps, tunnels, and in a macabre touch of theater, a "dead" Viet Cong sniper in a tree.

At the end of the exposition, Bob Cullum announced that a multi-million dollar improvements program was being planned for Fair Park which would include a new sports stadium, expanded parking and extensive landscaping that would lace the park with a network of canals.

The 1966 fair had pushed the attendance record closer to the three million mark. Overall income was up, but so were ordinary operating costs. Above that, there were out-of-the-ordinary expenses. Redevelopment required money, even when it was only in the planning stages, and the fair had invested heavily in studies and consulting fees. The Cotton Bowl needed new lights to satisfy requirements for color telecasting. The price tag for installing an ice rink in the Coliseum had gone up, and while there was no major league hockey franchise presently available for Dallas, the Chicago Blackhawks were ready to sign an agreement to locate a farm team in the facility.

1967

The fair reacted much as any business or household might when faced with pressing needs or immediate opportunities — it borrowed money. Insufficient parking space was another problem, but the solution was clearly beyond the fair's financial means. It was suggested that the individual directors lobby to have this item incorporated in the upcoming city bond issue.

Cullum had made a significant move to enhance the fair's influence by restructuring its governing body. On his recommendation, an honorary board was created for directors who had reached the age of 70. In January of 1967, nearly 40% of the State Fair Board became honorary directors, opening the way for a cadre of younger men to fill these vacancies. The new members included such key civic leaders as Lloyd Bowles and Dewey Presley, who would become actively involved in the organization, and two future State Fair presidents: Alfred I. Davies, Southwest Regional Manager for Sears, and Joe M. Dealey, president of the *Dallas Morning News.*

Cullum also went on record as favoring a new sports stadium for Fair Park. The suddenly successful Dallas Cowboys wanted a new home, and the fair president hoped to have this $20 million project as part of the bond program, although separate from the parking proposal.

Marching bands, floats and equestrian units assembled each evening for a twilight parade through the park.

Discarded food, containers and wrappers created a major problem for clean-up crews on the midway.

When the Dallas City Council ordered its priorities for the historic $175 million Crossroads Bond Program, $12.6 million was allocated for Fair Park. Of this amount, $2.8 million was designated for acquisition of property south of Pennsylvania Avenue to be used for parking, $2.5 million was directed for Cotton Bowl remodeling, and the remainder was intended to cover renovation of the Music Hall, air conditioning of the Coliseum plus landscaping and lighting improvements.

General manager Joe Rucker and his staff had been working over a period of time to improve the appearance of the park during the fair. Adding lights, fountains and flowers to the front of the grounds required only money; cleaning up the midway called for managerial muscle. The concessionaires, a tightly-knit group, some with histories of Fair Park operation dating back to the 1920s, resisted change. As independent business owners, they did not want to be told what to do, much less that they would have to spend their own money to do it. The midway of the early 1960s had a "used car lot" look — lean-to sheds, gaudy banners, peeling paint, hand-lettered signs, and by the end of each day, streets that were ankle-deep in trash. The sweet-sour smell that permeated the amusement and food areas grew progressively ranker toward the end of the run.

Rucker invaded this nether world with a five-year program of trash control, color schemes, professional signage and redesigned stands. Cooperation was assured when the first concessionaires to conform to the new standards reported doubling or tripling their grosses. The rest fell in line, though a few oldtime operators grumbled that they would miss the sound of beer cans being kicked. By the opening of the 1967 fair, 80 new stands were in place.

The Esplanade at night.

"Texas/International" was the exposition theme, though it proved difficult to secure exhibits sponsored by foreign countries. Funds for this purpose were drying up as governments increasingly relied upon commercial interests to tap the potent Texas market.

Rain cancelled the opening day parade for the first time in memory. Unfortunately, there was no rain the next weekend when police could have used help from the heavens in quashing a downtown brawl on Friday night before the big game. Upwards of 450 arrests were made for drunkenness and fighting; another 60 trouble-makers were turned over to juvenile authorites. A special police report indicated that only 22 of those jailed could be identified as students of either the University of Texas or the University of Oklahoma.

The Tahiti Nui Dancers entertained large crowds from a new stage erected in the middle of the Esplanade. Visitors also enjoyed Herb Alpert and the Tijuana Brass in two Cotton Bowl shows and Celeste Holm as "Mame" in the Music Hall. Ice Capades, a State Fair tradition since 1941, bowed out. More than $300,000 had been spent to remodel the Coliseum as the park's improved ice facility, but horse shows and livestock events required dirt flooring at fairtime.

Racial integration of midway attractions passed another milestone in 1967. Three white dancers appeared in the chorus line of the formerly all-black Cotton Club Revue.

1968

No one had called them America's team — yet, but the Dallas Cowboys twice had come close to upsetting the mighty Green Bay Packers in championship games, and America fell in love with these white knight underdogs. The "David vs. Goliath" saga of the Cowboys had given Dallas an excuse to cheer for Dallas again. The team became a rallying point which, perhaps as much as any single factor, helped the city emerge from the shadow of the Kennedy assassination.

The Cowboys' owner, Clint Murchison, Jr., wanted a new stadium. The issue was not simply the age of the Cotton Bowl or concerns about neighborhood crime. Control of stadium operations and division of revenues were the primary points of contention between Murchison and the State Fair of Texas. The Cotton Bowl's allegiance was to college football, by tradition and by reason of the close ties which existed between fair management and the SMU athletic department. The Cowboys were still the new kids on the block, and the Cotton Bowl was still the only stadium in town.

Murchison's demands appeared to be excessive. But when he couldn't get what he wanted from the City of Dallas, he struck a deal with the City of Irving. In late 1967, the Cowboys' owner announced he would move his team into a new $18 million sports facility to be built on a 90-acre site at the intersection of Loop 12 and Highway 183 — two-thirds of a mile outside the Dallas city limits.

People immediately chose sides. Football fans were either elated or outraged. Dallas city officials were indignant. Reaction among Irving residents was mixed. Blackie Sherrod, dean of local sports writers, described what followed "like a dual between man and alligator . . . vigorous but silent. Occasionally the combatants would rise to the top of the water and let out a gush of air, then dive back into their submerged struggle."

The State Fair went ahead with improvements for the Cotton Bowl that had been funded by the Crossroads Bond Program. Theater-type seating, restrooms, concession stands and drinking fountains were installed, and Republic National Bank sponsored the addition of modern scoreboards in each end zone.

Acquisition of land for parking, also part of the bond program, began in August, accompanied by charges of insensitivity to the needs of black residents in the neighborhood. And though this was essentially a battle between the city and the property owners over proceedings and indemnity payments, the fair, which had recommended that the expansion be made on the opposite side of the park toward Thornton Freeway, got caught in the crossfire.

There was irony in the selection "America '68" for the annual exposition theme. America 1968 had been a disaster: the Tet offensive, the assassinations of Martin Luther King, Jr. and Robert Kennedy, violent clashes between anti-war

Competition has always been the cornerstone of the State Fair. (above) Judges scrutinize hundreds of entries to determine prize winners in each division.

(right) Youngsters prepare animals they have raised for the Junior Livestock Show during the second week of the fair.

protesters and Chicago police during the Democratic Convention, even San Antonio's effort at producing a World's Fair had gone badly. And at that point, it certainly hadn't been a good year for the State Fair of Texas.

In a perceptive article on opening day, a *Dallas Morning News* reporter commented: "Trying to dissect the fair's appeal is, in fact, a little like cutting a hole in a balloon to see how it works . . . A fair is like a mountain. People come because it's there — all of it — and not necesarily because of any specific theme or attraction."

In 1968, sun shined on the new Superslide and Spelunker rides. It warmed the mini-skirted mob and their sockless male counterparts who packed the grounds on High School Day. It brightened the special outing for hundreds of crippled children, who were assisted by a corps of volunteers from the Dallas Fire Department.

Fairtime celebrities included Mrs. Frances Humphrey Howard, sister of the Democratic presidential nominee. Mrs. Howard was scheduled for an appearance in the Women's Building, and when a florist was late, anxious staff members stripped a nearby window box and presented the guest of honor with non-traditional, short-stemmed "Yellow Mums of Texas."

Exhibits covered everything from the Swedish guitar that Julie Andrews played in "The Sound of Music" to a popular series of 28-minute films entitled "Sermons from Science."

The Society for the Preservation of Pigtails sponsored a contest. Another competitive event was the World Bull Armadillo Championship. The winner, a fiesty nine-pounder named Brussel's Sprout, was facetiously auctioned off to the Chili Appreciation Society. According to the high bidder, the champion's future included a trip to the Terlingua Chili Cook-off, where he would become the main ingredient in the world's most expensive pot of chili.

Neon at night on the midway.

Midway through the final day, the State Fair of Texas broke the attendance equivalent of the four-minute mile. Bob Cullum thought about the three million visitors and grinned, "I don't see how we can make it a great deal bigger. But we do want to make it better."

1969

The big question, of course, was how to make it better, and when, and which plan to follow. The controversies of 1968 became the controversies of 1969. Homeowners affected by the parking expansion complained of unfair treatment. They received vigorous support from a Park

Cities church paired with the Fair Park neighborhood in a new community relations program called Block Partnership.

In February, Stanley Marcus spoke out at a meeting of the Museum of Fine Arts board suggesting that a new facility be built outside Fair Park. In March, the Wax Museum revealed that it too was considering a move.

With both the park department and the fair under fire, a rift developed between these two partners over the master plan for Fair Park redevelopment. According to park director L. B. Houston, there wasn't one. "We've got stacks of them," countered an exasperated Bob Cullum.

Since it was city property and city bond money at issue, there was little doubt which side would have the last word. The Park Board appointed a new committee to begin another study.

Meanwhile the fair, weary of criticism, in need of revenue and under pressure from the Park Board to schedule some off-season activities, announced it would introduce an April entertainment event. Spring Jubilee was planned as a mini-fair in conjuction with the annual Dallas Flower and Garden Show. The nine-day festival would be informal and inexpensive, built around family activities and special interest shows. On paper, the jubilee would build attendance for the flower show and vice versa.

The flower show was co-hosted by the fair, the Dallas Garden Center and the Dallas Council of Garden Clubs. Under an existing contract, two of the sponsors contributed volunteers and enthusiasm, the third stood the expense of producing the show. A charity dinner dance, held amid the floral decorations the night before the event, was a profitable fund-raiser for the Garden Center. The fair, however, depended solely on show admission income to recoup its costs.

Based on losses incurred over the years, fair officials had good reason to look for a way to build attendance at the flower show. Unfortunately, Spring Jubilee wasn't the answer. It succeeded only in making a bad situation worse. The festival would be rethought and redesigned four times, but it never stirred more than a ripple of public interest.

Youthful dancers on Texas-Mexico Day.

On July 20, 1969, Neil Armstrong climaxed a decade of American space exploration by taking a "giant step for mankind" on the lunar surface.

"Moon Year Exposition" was the logical theme choice for the State Fair. The keynote exhibit featured a one-third scale mockup of the Apollo command module and a walk-through display that told NASA's story from the earlier Mercury and Gemini programs to the moon landing. The theme show was housed in the New Dimensions Pavilion, a space-age name for the Varied Industries Building.

Visitors paid $1.50 at the park entry gates, the first price increase in 10 years. An admission ticket entitled the bearer to a day-long look at the sublime, the ridiculous, the good, the bad, and in some instances, the ugly.

A $19,600 Rolls Royce Silver Shadow glistened on a turntable, and the local dealer reported selling at least one of these costly models every month, establishing Dallas as a prime market for European luxury cars. In another building, an assortment of artifacts, purportedly salvaged from a sunken Spanish galleon, was showcased. Among the bells, crosses and doubloons was an emerald ring valued, according to its owner, at $250,000. This piece, the only uninsured item in the entire collection, disappeared at the end of the fair, touching off a lengthy police and FBI investigation.

One of the many champions at the 1969 State Fair of Texas was 87-year-old Mrs. Georgia Crockett, who took top honors with a blue and white afghan. It was Mrs. Crockett's second blue ribbon. She won her first for a quilt entered in the children's division in 1889.

A record 130 entries arrived for the annual twins contest sponsored by the women's department, but the fair's most celebrated set of matched siblings were Calvin and Coolidge, the first surviving twin calves ever born to a Santa Gertrudis cow. The identical duo tipped the scales at 4,600 pounds.

A new favorite was gaining ground in the snack food sweepstakes. Five years earlier, people had breezed past corn-on-the-cob stands looking for more traditional fare. But in 1969, 115,690 golden ears, skewered on sticks and dripping in butter, were sold to hungry fairgoers — enough, as calculated by an unappointed trivia statistician, to form an unbroken solid yellow line from the front gate to the Richardson city limits.

By 1969, hot buttered corn was on its way to becoming another State Fair success story.

1970

Astroturf was on the State Fair's shopping list in 1970. For $400,000, the Cotton Bowl was carpeted, and the lame duck Cowboys, waiting for completion of Texas Stadium, played another season in Fair Park.

The H. D. Lee Company offered to provide a new wardrobe for Big Tex's 19th birthday. The made-to-order outfit was shipped to Dallas and delivered to the park. But in an unguarded moment, Tex's 150-pound shirt, still in its box, was stolen out of a pickup truck. While police searched for the pranksters, the Lee company worked overtime to stitch up another size 90 garment, and Tex was buttoned into his new plaid top just nine hours before the gates opened.

Texas' own Miss America, Phyllis George of Denton, was on hand for the ribbon cutting. The State Fair joined forces with the department of oceanography at Texas A&M

The Berlin Pavilion in 1970.

University to produce "Oceanus," the theme feature of "Exposition of the Seven Seas." A 30,000 square foot pavilion was devoted to displays of undersea drilling rigs, submarine mockups, studies of marine life, maps and a simulated walk on the ocean floor.

Other noteworthy exhibits included a sampling of moon rocks collected by the astronaut crew of Apollo 11, and the modernistic Berlin Pavilion, a touring show presented in the United States as a goodwill gesture by the people of Berlin. The patriotic musical "1776" was featured at the Music Hall, and taken all together, the 1970 State Fair illustrated the organization's conscious effort to upgrade its program quality. The fair had moved far beyond the realm of hot dogs and carnival rides, though concession revenues still paid the educational and cultural tab.

As was frequently the case, the fair provided a forum for political views. Blacks protested South Africa Day. Another organized group brought petitions to the grounds and collected over 100,000 signatures calling for the quick release of Vietnam POW's.

The Texas Department of Agriculture organized a new consumer-oriented show about Texas foods and fibers in a building which the fair modernized and redecorated for this purpose. Southwestern Bell's exhibit was built around a frontier town where kids picked up special phone lines to eavesdrop on conversations between Disney characters.

The 1970 exposition produced its share of heartwarming stories such as the 134 ribbons won in cooking contests by the combined branches of the Joe E. Patek family. But it also formed a backdrop for human tragedy when another Saturday night swimmer got past security and drowned in the lagoon.

Despite the cold and a collage of negative headlines, total attendance passed three million for the second time. Fair president Bob Cullum took pride in the numbers, but he also acknowledged a need for new voices to be heard in shaping programs for the decade ahead. To this end, Cullum appointed

Quilts have captured the fancy of fairgoers since the exposition's earliest days.

One of 1970's more prestigious attractions was the annual Symphony Spectacular presented in the Cotton Bowl by the Dallas Symphony Orchestra, Dallas Civic Ballet and U.S. Fourth Army Band. Financed by the fair, these popular concerts, the Dallas Symphony's first before mass audiences, traditionally concluded with the "1812 Overture" dramatized by cannonading on the field and the burning of Moscow in fireworks. The original show date was rained out. On Sunday, October 18, the various performing groups assembled to try again. The evening was cool and dry; a small, enthusiastic crowd waited. But the Symphony's union contract stipulated that the orchestra did not have to play if the temperature dropped below 68 degrees, and the thermometer read 59. The members took a vote, packed up their instruments, and despite pleas from their conductor and fair officials, amid hisses and jeers, they walked off. The Ballet and Army Band went on with the show and received a prolonged standing ovation.

An orchestra spokesman said later, "It is the professional prerogative of any professional artist or group of artists to decide when conditions will prevent their presenting an artistically acceptable performance." But *Dallas Times Herald* cartoonist Bob Taylor expressed the public's reaction and took aim at future fund drives when he drew a Dallas citizen ignoring the Symphony's tin cup with the comment: "Sorry, but when the temperature gets below 68 degrees, my wallet won't perform."

12 young businessmen to an advisory task force. Cullum expected a new generation of leaders to emerge from this group, and in late 1972, two task force members, attorney Russell B. Smith and Roosevelt Johnson, Jr., director of the Dallas Urban League, would be tapped to fill vacancies on the State Fair Board — Johnson thus becoming the first black to serve in this capacity.

1971

Though racial tensions certainly existed, Dallas had escaped the violence that plagued many large cities during the 1960s. Credit belonged to reasoned leadership in the black community and to the sometimes-maligned Dallas establishment, those bankers and businessmen in a position to dictate that barriers be removed.

Integration of Dallas schools began one grade at a time, a "stairstep" plan that moved too slowly to satisfy federal authorities. Immediate desegregation then was ordered at all grade levels, though except for the city's few racially-mixed residential areas, compliance had minimal impact. Dr. Emmett Conrad became the first black member of the Dallas School Board in 1967, and DISD faculties were integrated in

1970; but it was court-mandated busing in 1971 that changed the demographic balance of the Dallas school system and gave rise to "white flight."

White flight also threatened Fair Park at this point. Whatever the official reasons offered, many organizations did not want to deal with the issue of sponsoring white-oriented events in a black neighborhood. This deeply concerned retiring mayor Erik Jonsson, who referred to the park as "Dallas' safety valve," the only place in the city where people of all races could gather for entertainment on a personal level.

Renovation of the Music Hall, approved by voters four years earlier, had been on hold, in part at least, until the recent furor over land acquisition cooled down. Some now questioned the wisdom of the project, arguing that it made more sense to put these funds toward construction of a new theater "in a more central location," a euphemistic phrase which ignored the fact that there was no more central and easily-accessible site anywhere in Dallas. Nor was much support for the Music Hall to be found in the black community. Here the contention was that South Dallas had greater needs than a performing arts center that few nearby residents would use. Despite the controversy, the Dallas City Council reaffirmed its commitment to Fair Park and approved plans for remodeling and modernizing the 46-year-old hall.

"All in the Family" made its television debut in 1971, while cigarette advertising disappeared from home screens. Movie audiences sniffled through "Love Story;" tennis star Billie Jean King became the first woman athlete to earn $100,000 in a single year; and after the longest trial in California history, Charles Manson and members of his so-called family were convicted for the orgy of murders in the home of actress Sharon Tate.

"EXPO/TRANS/PORT," the 1971 State Fair theme, celebrated the latest developments in transportation including supersonic planes and pollution-free automobiles. A major attraction in the pavilion was a model of the Dallas/Fort Worth Regional Airport under construction midway between the two cities.

Musical entertainment choices ranged from "Promises, Promises," the final show to play the "old" Music Hall, to Grand Funk Railroad, to jazz trumpeter Al Hirt to local favorite Jesse Lopez.

The Texas Limousin Association spotlighted the first purebred Limousin bull ever shown in the state, and the International Federation of Charolais Breeders hosted its eighth convention, the first to be held in North America.

Familiar faces among the fairgoers included actor Yul Brynner and his new French wife; Monaco's royal couple, Prince Rainier and Princess Grace; and fabled oilman H.L. Hunt, who was pushing HLH products and patriotism from a

State Fair president Bob Cullum converses with Princess Grace in 1971.

booth in the Women's Building. The 82-year-old billionáire, known for his frugality, worked an eight-hour shift daily. On one occasion when he forgot his exhibitor's pass, Hunt waited patiently for someone to come to the Grand Avenue Gate and identify him, rather than pay the standard $1.50 admission.

Missing faces in 1971 were those of the Dallas Cowboys. The team played its final game in the Cotton Bowl against the Washington Redskins the Sunday before the fair opened.

Total attendance reached 3,134,646, and of that number, perhaps no one enjoyed the event more than 19-year old James Rogers of South Dallas, who followed the clues and found Colgate's mystery man for a $4,000 pay-off on a 65-cent soap investment.

The defection of the Cowboys indirectly led the fair to spend money it really didn't have on a purchase it would ultimately regret. The owners of the Skyride had been offered a good price to move the aerial tramway to Mexico City. Sensitive about the possibility of losing another popular attraciton, the State Fair borrowed $300,000 and bought the ride. Over the years, the fair had steered clear of outright ownership of rides. But in the case of the Skyride, and later the antique carousel and Comet Coaster, the organization stepped in to preserve the attraction for Fair Park.

Shortly before Christmas, construction crews began a $5.4 million facelifting program for the Music Hall. Included in the renovation were three new wings to accommodate concourses, a stage annex and a spacious intermission lounge and restaurant which extended outward from what had been the front of the auditorium. The main entrance was shifted to the south side facing the parking lot. Dressing rooms and rehearsal facilities were remodeled and expanded. New seating, lighting, air conditioning and a totally redesigned sound system were elements in the architectural package, which won awards for the firm of Jarvis, Putty, Jarvis even before the theater reopened.

Construction workers build a new wing onto the original arched entry to the Music Hall.

1972

The Music Hall was only part of an extensive structural restoration program carried out in 1972. The Band Shell received needed attention — paint and repairs for the stage and backstage areas plus new lights and seating. By summer, the improved amphitheater was in shape to host a five-week, 15-concert pops series presented by the Dallas Symphony Orchestra, and a one-performance "Evening of Shakespeare," which laid the foundation for an annual Shakespeare Festival in the park.

An overnight fire destroyed the old Poultry Building and left an ugly gap on the east side of the agricultural complex. To fill the space, general manager Joe Rucker designed the Texas Crafts Village, a hideaway of stalls around an enclosed patio for folk art displays and entertainment.

Wax World recreated scenes from the lives of American presidents.

Wax World, a themed museum highlighting scenes from the everyday life of American presidents, moved into the area vacated by the Southwestern Historical Wax Museum. The $400,000 project, under the direction of Peter Wolf and a new company called Leisure and Recreation Concepts, featured animation and quality reproductions, but the public would never show much interest in Harry Truman's piano-playing or John Quincy Adams' skinny-dipping.

In 1972, the Electric Building tower was stripped of its clock, lettering and identity. Renamed the Better Living Center, the structure looked as it had in 1936, distinguished by a white column with gold fluting and an eagle on the top. Throughout the park, fountains, floors, furniture — even the glass brick walls in the old Margo Jones Theater, were restored to an approximation of their Centennial splendor, a painstaking and costly process appreciated in preservationist circles, but with little direct benefit for the public. If nothing

The 1972 theme pavilion spotlighted Bernice Kent's collection of '30s memorabilia.

else, it was an ambitious and questionable undertaking for the State Fair at a time when the organization was stretched financially.

With a continuing bent for nostalgia, fair officials chose "The Dazzling Thirties" for the 1972 theme. A few eyebrows were raised. Some observers commented that not everyone who remembered the decade would describe it as dazzling, but as J. T. Trezevant had noted 80 years earlier, people came to the fair to be entertained, and the '30s theme was potentially far more entertaining than the scientific and technological focus of the three preceding expositions.

The New Dimensions Pavilion recreated an era through artifacts, photos, posters, news reels and a special show "Expositions of the 1930s" mounted by the Dallas Museum of Fine Arts. The Golden Age Film Festival offered movie buffs a program of classics from "Grand Hotel" and "Citizen Kane" to "A Day at the Races." The Dazzling Thirties Revue in the Band Shell took audiences on a sentimental journey with the Ink Spots, sensational tap dancer Gene Bell, the Johnny "Scat" Davis Orchestra and an entire troupe of resurrected talents.

"No, No, Nanette" reopened the Music Hall. If the show was not vintage 1930s, its stars, Don Ameche and Evelyn Keyes, definitely were. In contrast, the theater was plush and contemporary, as luxurious and elegant as its steadfast champion, Dallas Park Board president Dr. William B. Dean, had promised.

Art Linkletter, who owed his big career break to the Texas Centennial, took part in the opening ceremonies. Another guest from those thrilling days of yesteryear was the Lone Ranger, Clayton Moore, who appeared at the Dodge display in the automobile show.

The Marlboro Chuckwagon attracted notice from fairgoers by combining authentic range-style cooking with western entertainment. In the Cotton Bowl, the Parade of Champions, an adjudicated competition for marching bands from Texas high schools, drew respectable crowds in its inaugural outing.

(below left) Broadcast personality Art Linkletter assists State Fair president Bob Cullum and Mayor and Mrs. Wes Wise with the ribbon cutting ceremony in 1972.

(below right) The Lone Ranger signs autographs for young fans.

The 1972 State Fair of Texas attracted three million visitors, brought in substantial revenue and garnered good reviews, but the organization's earnings again failed to keep up with its expenditures and obligations.

General manager Joe Rucker was a gifted individual whose misfortune was to be in the right job at the wrong time. Rucker had done yeoman work on the redevelopment plans of the early 1960s, succinctly analyzing future needs of the fair and the park, advancing ideas that in some instances required 25 years to get due consideration from the decision-makers at City Hall. Not all of Rucker's proposals were realistic, however. He was admittedly more a visionary than a practical businessman, a perfectionist in the execution of programs, rather than an executive closely attuned to the bottom line. Circumstances beyond anyone's control had triggered the fair's slide into a mire of red ink, but despite good crowds and innovative entertainment, the organization was sinking deeper, and measures intended to turn the situation around seemed only to compound its seriousness.

The board went outside the organization and outside the downtown business community to hire a new general manager with a proven track record in the amusement industry. Wayne H. Gallagher, formerly vice president of administration for Six Flags, Inc., was charged with the task of putting Humpty-Dumpty's finances back together again.

A Blue-Ribbon Business

1973-1984

1973

Richard M. Nixon had been elected to a second presidential term by the largest popular majority in the nation's history. But even as Nixon took the oath of office, the stage was being set for a twisting, turning, emotional year filled with tension, drama and elements of national despair. In the court of Federal Judge John J. Sirica, a trial had begun for the seven men accused in a bungled burglary six months earlier at the Watergate headquarters of the Democratic National Committee.

On January 22, 1973, former president Lyndon B. Johnson died of a heart attack. Five days later, a cease fire accord was signed to end the hostilities in Vietnam.

By summer, televised committee hearings, grand jury indictments and resignations by top officials had shaken public confidence in the Nixon administration, and there was no sign that the political tremors had run their course.

Despite, or perhaps because of the ongoing government crisis, the outdoor amusement industry was enjoying a banner year, and new general manager Wayne Gallagher predicted that Watergate would have a positive effect on attendance at the 1973 State Fair of Texas, if the weather cooperated.

Wayne H. Gallagher, General Manager, 1973-

The theme, "World Gateways Exposition," was tied to the imminent opening of the Dallas/Fort Worth Regional Airport. In one of his first moves to right the fair's finances, Gallagher replaced the theme pavilion with a revenue-producing exhibit building. The redesigned facility was called the Embarcadero after the wharf marketplace in San Francisco, and though few fairgoers knew its origin, and many had trouble pronouncing it, the name stuck, possibly because it avoided precise definition.

The international focus of the 1973 fair centered around an elegant cultural display from Japan, highlighted by daily presentations of a 500-year-old tea ceremony, and precedent-making participation by Romania which sponsored its first-ever trade exhibit in Texas. Unfortunately, a small South African booth received most of the attention. The State Fair traditionally invited every nation recognized by the United States government to take part in the annual exposition. South Africa had sponsored displays before, but after objections were raised in 1970, the fair, in deference to strong sentiment against apartheid policies, agreed not to schedule a South Africa Day. Three years later, after some turnover in personnel, a new staff member inadvertently included South Africa on a list of special days in an advance promotional piece. Outrage was immediate. Even before the gates opened, activists demanded that the fair cancel something it hadn't scheduled. Explanations were refused, and 10 protesters showed up on the first day carrying signs that read: "Texas State Fair Supports Neo-Slavery." The non-issue lingered for days. Finally the fair made an official statement that there was no South Africa Day, and the demonstrations stopped.

Feminists made their presence known at the fair, not by protesting, but by entering cooking contests. Members of the National Organization for Women won 12 ribbons in the first week and announced that their cash awards — a total of $25 — would go to support the Equal Rights Amendment.

Other contestants in other categories included former governor John Connally, whose Santa Gertrudis bull earned the top prize for the breed, and current governor Dolph Briscoe and presidential counselor Anne Armstrong, who brought entries to the Pan American Livestock Exposition.

With no strong touring shows available, Music Hall management reached back into the theater's storied past to repeat the original 1925 attraction, "The Student Prince." But 48 years had dimmed the Romberg magic, and this revival joined the miniscule group of fairtime musicals to lose money.

A good final weekend, after a scourge of rain and cold, boosted attendance and income figures to very respectable,

(above) German Day at the fair translates into lots of music, dancing, wurst and beer. Other special days have honored Texans of Czech, Norwegian, Scotch, Mexican and Turkish heritage.

(right) Former governor John Connally is actively involved in the annual Santa Gertrudis cattle shows.

though not record levels. The *Dallas Morning News* published an in-depth business analysis of the event and concluded that "next to Christmas," the fair was the best thing that happened to the local economy each year.

1974

The resignation of Vice President Spiro T. Agnew, Nixon's "Saturday Night Massacre" firings and the whole mosaic of illegal acts and power abuses known as Watergate weighed heavily on the country. Soaring prices and worldwide energy shortages contributed to a generally gloomy prognosis for 1974.

In North Texas, however, the long-awaited unveiling of D/FW, the world's largest airport, kindled enthusiasm and pride. The 17,800-acre site, an insular world of earth-toned concrete and smoky glass connected by a futuristic network of roadways and spiraling ramps, straddled the Dallas-Tarrant County line. The project was a remarkable cooperative venture between two cities that chose to bury historic rivalry in order to serve mutual interest.

D magazine appeared on local newstands, and the lead article in the first issue named State Fair principles Bob Cullum, John Stemmons and James Aston as the three most powerful men in Dallas.

Bulldozers returned to Fair Park in 1974. The State Fair, after requiring a loan transfusion to cover operating expenses the previous year, was on the road to financial recovery. The organization had caught up with its obligations and even had a small amount to invest in the first stages of a midway improvement program. Using a partial pre-payment of the games concession contract, the fair built a new pedestrian mall featuring trees, brick and aggregate walkways and special lighting between Big Tex Circle and the plaza at the main entrance to the Cotton Bowl. Another construction project involved razing the old roller rink to make room for six additional rides, and the city allocated community development funds to redesign the food service area on Cotton Bowl Plaza.

"Our long national nightmare is over," assured Gerald R. Ford on August 9, 1974, when he was sworn in as the 38th president of the United States replacing the departed and disgraced Richard Nixon. The former congressional leader, a gregarious, uncomplicated man, moved to heal the country's shattered morale.

The State Fair of Texas added rodeo to its lineup of attractions and themed the 1974 show, "Exposition of the

(top) "Exposition of the West" marked the return of rodeo to the lineup of State Fair attractions.

(below) In 1974, Spain sponsored a major exhibit of commerce and culture which featured daily performances by a troupe of flamenco dancers.

West." The Coliseum had hosted the National Finals Rodeo from 1959-1961, as well as winter and spring rodeos from 1962-1967, but historians had to dig back 40 years to find this quintessentially western sport at fairtime. Management believed a new audience existed and hired Austin's Tommy Steiner to produce the nine-day event.

In other changes, Gallagher announced that the price of an adult ticket would go to $2.00 and the exposition would open a half-day earlier than usual, affording early-birds an opportunity to preview the show and giving the fair a chance to see that everything was in place and working before the Saturday legions descended.

One of the opening day features was a gridrion contest between Grambling and Prairie View. It was advertised as "The Game and a Half," calling attention to a halftime show featuring the schools' superb marching bands. The half proved more entertaining than the game, which Grambling won 61-0.

Debbie Reynolds starred in the title role as the little Irish piano tuner in "Irene," with a strong supporting cast that included Patsy Kelly, Ruth Warrick and Hans Conreid. The musical went on to break the Hall's all-time gross receipts record, and Bob Cullum charmed Miss Reynolds by giving her a pendant fashioned from a 1925 gold piece to commemorate the occasion.

The high point of the inaugural State Fair Rodeo saw 21-year-old Donnie Gay of Mesquite, a contender for world champion bull riding honors, ride the previously unrideable #104 for the requisite eight seconds. Another western-clad visitor was Michael Haynes, the suave and apparently unforgettable Winchester Man of TV commercial fame. R. J. Reynolds had pulled the ads for its little cigars more than a year earlier, but Haynes was mobbed wherever he appeared on the grounds.

Smiles were abundant at the end of the run. The 1974 exposition set attendance marks and earned a handsome profit, giving the fair some cash to work with as it prepared a star-spangled 1975 edition which would get the jump on the nation's Bicentennial celebrations.

1975

The City of Dallas used bond money to build a modern Livestock Judging Pavilion in 1975. The clean, well-lighted arena measured 100′ x 244′ and provided seating for 1,200 spectators. A new 379-stall, open-sided cattle building adjoined the pavilion.

With the Astroturf paid off, the State Fair channeled funds into further development in the Cotton Bowl Plaza area and bit the bullet for a $90,000 rehabilitation of the tower on the Better Living Center. Construction was by no means the only expensive item on the fair's budget. Visitors had noticed and appreciated improvement in the overall cleanliness of the park, but the price of progress in trash management had risen to $200,000 for 17 days, a 400% increase over the amounts spent for this purpose in the early 1970s.

The fair installed a central ticket system on the midway prior to the 1975 exposition. Long lines and cash transactions at individual rides were eliminated by locating 12 colorful kiosks around the amusement area. Patrons purchased coupons at these booths and exchanged a specified number for admission to each attraction. The system also provided an effective control in the accounting of ride revenues.

Big Tex set the tone and style for "The Yankeedoodle Dandy" State Fair of Texas. Tex sported a new candy-striped western shirt and stars on his boots.

The exposition marched to a bicentennial beat in 1975 from flags and streamers flying throughout the park to an animated diorama and scaled reproduction of Philadelphia's Independence Hall in the Food and Fiber Pavilion. George Zambelli produced a nightly fireworks show.

Each of the museums keyed special exhibits to the theme, and the rotunda of the Better Living Center was transformed into a gallery for oil paintings of American presidents and first ladies. The historical pageant, "I Am Old Glory," was presented in the Cotton Bowl.

For all the patriotic emphasis, the star performers of 1975 wore costumes of brilliant pinks and greens studded with sequins and feathers. They played "pans" and jump-danced to the lively, Latin-influenced melodies of calypso. After months of negotiation and an inspection of the grounds by high-ranking government officials, the Caribbean island nation of Trinidad and Tobago sent its finest steelband to the fair. Texans were enchanted and amazed at the full orchestral sound produced by instruments forged from oil drums.

(below left) The State Fair's antique merry-go-round boasts 66 wooden horses crafted by the acknowledged master of carousel carving, M.C. Illions, in the early 1920s.

(below right) The jump dancers and champion steelband from Trinidad and Tobago created a sensation at the 1975 State Fair.

Country star Charley Pride shared top billing with Waylon Jennings for a Cotton Bowl concert.

With the variety of entertainment choices offered to fair visitors each year, it was inevitable that the popularity of certain attractions doomed the success of others. The Salesmanship Club promoted two Cotton Bowl concerts, the first starring the soft rock duo of Seals and Croft, the other featuring Charley Pride and Waylon Jennings. But true to form, fairgoers virtually ignored these reasonably-priced shows and opted for the acres of free entertainment outside the stadium. The Music Hall tried a split-engagement format: one week for Juliet Prowse and one week for Johnny Cash, with disappointing results at the box office.

But there were far more winners than losers in 1975. Attendance reached 3,176,028, and the fair greeted the new year with restored financial vigor, a spectacular turn-around accomplished in just three years.

1976

The State Fair had guessed correctly in celebrating the nation's 200th birthday a year early. Americans stood in line to see historic documents and memorabilia on the Bicentennial Train that toured the country, and millions watched the parade of tall ships sailing down the Hudson River, but by fall people were surfeited with patriotism and pageantry. "October Magic," theme of the 1976 fair, presented a welcome alternative.

Although the fair's staff was accustomed to setting up for the advent of three million people, preparations for just one 1976 guest created unexpected challenges. President Gerald Ford decided to attend the fair on opening day. Since it was an election year, Ford's political advisors wanted the president

The United States Marine Drum and Bugle Corps from Washington, D.C., is a perennial fairtime favorite.

seen by as many potential voters as often as possible. This translated into a ride in the downtown parade, a tour of the livestock facilities, participation in the ribbon-cutting ceremonies, a formal luncheon and a coin-tossing appearance at the Texas-OU game.

Secret Service personnel and multitudes of Republican functionaries invaded the park in September. Security precautions were understandably complex. Certainly no one had forgotten the last time a president rode through downtown Dallas in an open car, and two attempts had been made on Ford's life the previous year.

But it was not simply the president, his assistants, the Secret Service, local Republican candidates and committee members, and three bus-loads of national and foreign press that arrived on opening weekend. The staff of NBC's "Today Show" moved in for a series of telecasts from the grounds; a film crew from "Semi-Tough" showed up to get crowd shots in the Cotton Bowl; George Plimpton appeared to personally supervise his gala international fireworks show scheduled in the park on Friday night; and fans from Norman and Austin hit town for a well-known football game.

The presidential visit went almost exactly as planned. Illness prevented Bob Cullum from hosting the welcome. Fair vice president John Stemmons filled in ably on short notice, but there was concern for Cullum, now in his tenth year as State Fair president, who enjoyed such occasions too much to miss one if he could possibly help it.

Three days later, the Democrats asked for equal time. Rosalynn Carter would be in Dallas on the next Saturday and wanted to visit the fair. Arrangements were made for her to deliver a short speech and serve as an honorary judge in the cake baking contest. The creative arts department requested and received Mrs. Carter's favorite recipe, but none of the blue-ribbon cooks was able to reproduce a moist peanut cake, even after six tries. The attractive wife of the future president didn't stay long enough to sample the final cardboard-textured effort. She presented awards to the winners, spoke briefly about her husband's good qualities, accepted a bouquet of yellow roses and was hurried into a waiting car, already late for the next stop on a long campaign trail.

Most fairgoers had minimal interest in the political sideshows. They attended the 1976 exposition to see John Raitt in "Shenandoah," or the new production of Dancing Waters staged in the center of the Esplanade, or the 130-carat "Light of Peace" diamond valued at $5 million and displayed on weekends under heavy guard by the Zale Corporation.

Currency problems in Latin America reduced participation in the Pan American Livestock Exposition, and world economic conditions decimated the ranks of foreign government sponsored exhibits. Overall attendance was off

President Gerald R. Ford (pictured with State Fair general manager Wayne Gallagher, Texas governor Dolph Briscoe and fair vice president John Stemmons) officially opened the 1976 exposition.

(above left) The fair purchased the Comet Coaster for $135,000 in 1976.

(above right) Dancing Waters — a spectacle of fountains, lighting and music — was presented from the center of the Esplanade.

about 5%, but spending remained high, leaving the fair with the funds it needed to continue midway redevelopment and work begun on the Comet roller coaster. The fair had purchased the giant ride for $135,000 the preceding spring and would spend another $165,000 in the first year of an ongoing renovation program.

1977

"It was a year of slow and undramatic adjustment to changing circumstances, as if people were pausing before some new burst of national energy," wrote *New York Times* columnist James Reston.

In 1977, Americans stayed home for eight consecutive nights riveted to television sets as the saga of "Roots" unfolded. Moviegoers found escape in the pulsating disco sequences of "Saturday Night Fever" and the "Star Wars" adventures from "a galaxy far far away."

Preparations for the State Fair of Texas emphasized an improved exhibits program. President Bob Cullum, fighting a battle for his health, found energy to contact leading food industry executives about participating in a special pavilion at the fair. The project, coordinated by staff member Jeanne Baker, gave rise to a theme, "The Great Food Round-Up." By opening day, the Tower Building, formerly the Better Living Center, was filled with an impressive array of state-of-the-art exhibits sponsored by names familiar to any grocery shopper: Kraft, Nabisco, Quaker Oats, Pepsi Cola, Frito-Lay, Heinz, Pillsbury, La Choy, Swift, General Electric and others. Much to the delight of fairgoers, the old time tradition of free samples was revived, and a literal feast of freebies was offered every day.

Chefs from top local restaurants demonstrated their culinary skills in the General Electric Kitchen Theater for "The Great Food Round-Up."

New food items sold at concession stands included Gulf-fresh shrimp, egg rolls and something called a Texas Grinder, a slab of sourdough filled with beef, cheese and spices, pressed in a waffle iron and dipped in melted butter. El Chico sponsored a Sunday chili cook-off in a picnic tent and surrounding area. The chili-heads had a marvelous afternoon, but management discovered this group had a tendency to spread their activities beyond prescribed limits. Eventually the simmering pots and noisy party monopolized a valuable parking lot, an unpardonable fairtime sin that guaranteed there would be no second annual chili cook-off the following year.

Trinidad and Tobago returned to the fair represented by the champion steel band, the winners of the island's Best Village competition and lovely Penny Commissiong, Miss Universe of 1977.

Janelle "Penny" Commissiong, Miss Universe of 1977, represented Trinidad and Tobago at the fair.

The Centennial Building, as the World Exhibits Center was now called, spotlighted merchandise from trampolines to Tibetan rugs, donut-making machines to log cabin homes, plus the familiar international bazaar for imported goods. The Embarcadero offered even more variety, including best-selling Elvis memorabilia. The rock star's death in August had spawned an enormous market for t-shirts, buttons, patches, posters and other paraphernalia.

Early on the Friday evening before the Texas-Oklahoma game, a police spokesman, assessing the downtown situation, commented, "I think we're beginning to see the end of this thing . . . there's just not as much interest as there used to be." But after 10 p.m., when local high school football games were over, the crowd grew, and by morning 244 had been arrested, the largest number in five years.

Johnny Rodriguez and Eddie Rabbitt each headlined Cotton Bowl shows, and Debbie Reynolds appeared in a revival of "Annie Get Your Gun." Bob Cullum renewed his friendship with the vivacious star and jokingly offered to encircle the gold pendant with diamonds if she broke her own top gross record set in 1974. She did, by just $3.50, making it

Action — on the field and off — characterizes the annual gridiron battle between traditional rivals Texas and Oklahoma. The game is played at the State Fair in Dallas — a neutral site equidistant from the Austin and Norman campuses.

an expensive promise that Cullum fulfilled. One of those in the audience the final weekend was Reynolds' daughter, Carrie Fisher, whose film career had skyrocketed in the year's biggest movie hit, "Star Wars." The 21-year-old actress toured the grounds for several hours, visiting midway shows, buying gifts for friends, sampling calorie-laden goodies, without a single fairgoer recognizing Princess Leia in everyday clothes.

Total attendance in 1977 was 3,178,455, a new and possibly enduring all-time record for 17 days. Revenues also surpassed all existing marks, and Dallas Mayor Robert Folsom suggested that Fair Park would make an ideal site for a Sesquicentennial Exposition in 1986.

1978

Inquiries were made as to the feasibility of having an official world's fair to commemorate Texas' 150th Birthday. But Knoxville had already been sanctioned for 1982, and New Orleans had received U.S. approval for 1984. It was therefore decided to pursue the idea of a major exposition with a regional focus.

By 1978, progress could be seen in the development of year-round park programming. Taking into account performances, rehearsals, graduation ceremonies and other community events, the Music Hall was occupied more than 300 dates out of each year. The popular two-day Artfest and three-week Shakespeare Festival commanded large followings. Museum exhibitions, school field trips, weekend special interest shows, livestock activities and park recreation programs contributed to an estimated five million annual visits by people using Fair Park. These were not pie-in-the-sky numbers, nevertheless a vast segment of the public persisted in looking at the park as a sleepy, deserted and probably dangerous place that came to life once a year for the State Fair. Part of the problem was the sheer size of the 250-acre facility and the diverse interests of the groups meeting there. Opera patrons at the Music Hall were generally unaware that an herbal workshop was taking place at the Garden Center or a rabbit show was going on in the Poultry Building. Another factor was the publicity generated by the Dallas Museum of Fine Art to obtain support for the public funding it needed to build a new downtown museum. This, when combined with talk of the Dallas Symphony wanting a concert hall of its own and rumors that SMU was considering moving its football games to Texas Stadium, prompted cynics to suggest that the last one leaving Fair Park ought to turn out the lights.

Former major league pitcher Dean Chance, a winner of baseball's coveted Cy Young Award, in his new career as an owner/operator of midway games.

Rock music fans jammed the Cotton Bowl on a scorching summer day in 1978 for the first-ever Texxas World Music Festival.

Apart from opinions and impressions, there was one obvious, undisputed fact. The largest and most expensive structure in the park sat empty most of the year. That would change in 1978. To the horror of gridiron traditionalists, the Cotton Bowl became known as the perfect venue for outdoor rock concerts. Of course concerts, per se, were not new to the Bowl, but these were festival shows — marathon, all-day, most-of-the-night celebrations featuring four or five major performing groups and attracting thousands of teenagers and young adults. Built into the summer festival format were concerns about heat, security, alcohol, drugs, traffic, noise and clean-up.

The first Texxas World Music Festival was an overly-ambitious three-day event on the first weekend in July. There were problems, but fewer than expected. The promoters, Pace Concerts of Houston, and the State Fair staff learned the basics of a new business, and the fair discovered that the role of landlord with a share of concession revenue could be an off-season gold mine.

A new Dallas City Hall designed by I. M. Pei was dedicated in 1978, but in a July bond election, voters surprised complacent civic leadership by approving only the "bread and butter" issues. Proposals relating to a new art museum and engineering studies for a Fair Park-downtown boulevard link were defeated.

"Celebration — Texas Style" was a fair weather fair, and 17 days of sunshine assured a successful run. Inflation boosted adult admission to $2.50, but the public was accustomed to prices going up everywhere in 1978.

The State Fair Rodeo, in its fifth year, had not developed into the standout attraction originally predicted. Weekend attendance was good, but Monday through Thursday performances did not draw well. The fair had subscribed to the

Fairgoers in the year 2050 will be able to examine the contents of a time capsule planted 100 years earlier by General Manager Jimmie Stewart at the Mid-Century Exposition.

Befitting an organization in the business of celebrations, the State Fair of Texas has trumpeted each of its anniversaries, using these milestones as an excuse for a little more hype and an extra dose of dazzle.

The fair, however, has proven itself better at giving parties than keeping track of numbers. Somewhere along the line confusion developed as to how many and which birthdays counted in reckoning an anniversary. Thus, the organization feted number 25 in 1910; number 40 in 1926, 16 years later; number 50 in 1938, after another 12 years; and number 75 in 1960, following a 22-year interval. The 100th anniversary is slated for 1986 — 26 years after the 75th.

By actual tabulation, during the month of October on the southeast Dallas acreage known as Fair Park, there have been: one Dallas State Fair (1886), 17 Texas State Fairs (1887-1903), one Fall Festival (1904), one year as a World War I military camp (1918), one year devoted to construction (1935), two years for special expositions (1936-1937), four years absence for World War II (1942-1945) and 73 editions of the State Fair of Texas.

But it's not polite to count the candles on a senior citizen's cake, and during the Texas Sesquicentennial year, the state's most durable entertainment institution will celebrate a century — 1886-1986.

idea of presenting rodeo strictly as a sports event, as opposed to rodeo structured around a name entertainer. In something of a philosophical compromise, the 1978 rodeo featured both a name and entertainment. The name belonged to Larry Mahan, six-time world champion cowboy, and Mahan and his Rambling Rodeo Revue furnished the entertainment.

Despite the ongoing emphasis on attendance, fair officials wished that one group of daily visitors would have stayed

(below left) Square dancers, gospel singers, dancing schools and choral organizations join with professional performers to provide a program of continuous entertainment at locations scattered throughout the 277-acre park.

(below right) The Tony Award-winning musical "A Chorus Line" attracted more than 68,000 people over a 24-performance run in 1978.

home in 1978. Members of the Hare Krishna religious sect wandered through the ground distributing small flowers, flags, even Big Tex buttons, and aggressively sought donations from anyone accepting these "gifts." After numerous complaints, the fair tried to restrict the group's activities to a booth, but the Krishnas went to court and secured an injunction permitting them to continue what they defined as a religious practice. The fair posted signs advising fairgoers of their right to ignore the Krishnas' requests, and Big Tex boomed out periodic warnings until someone recognized that this was really not Tex's kind of message.

"A Chorus Line," one of the brightest new Broadway shows in years, dazzled critics. Fairtime audiences demonstrated enough sophistication to handle adult themes and language, something that wouldn't have surprised Corinne the Apple Dancer.

1979

After 12 years as president of the State Fair of Texas, Bob Cullum retired, and the board filled its top post by electing Alfred I. Davies, a retired Sears executive who had been active in the development of the Junior Livestock auction sale program.

Cullum, somewhat like Gaston years before him, had taken charge at a particularly stormy point in the fair's history and had guided the organization through financial shoals, weathered criticism, changed course when necessary and repaired damage as seamlessly as possible. By 1979, the State Fair was operating on a firm economic foundation with a strong management team in place. Fairtime attendance seemed solidly entrenched near the three million mark. The

Alfred I. Davies, State Fair President, 1979-1982.

Repeated surveys have confirmed that, year after year, the automobile show is the fair's most popular attraction.

Musical production numbers are combined with a look at fall and holiday fashions in daily shows produced by the creative arts department (the name assumed by the women's department in 1975 to describe its broader range of activities).

exodus of organizations leaving the park was nearly complete, those wanting out were either gone or had given notice.

One change that Cullum had resisted took place quickly under his successor. When a vacancy opened on the board, the directory elected its first woman member — the doyenne of Dallas realtors, Ebby Halliday.

After a successful season of summer concerts in the Cotton Bowl, the synthetic turf was replaced, and the fair got down to the serious business of preparing for its 1979 exposition, "Salute to Good Neighbors." The theme had been selected after a large exhibit from Canada was assured and with the expectation that the government of Mexico could be persuaded to participate in a comparable fashion. But that commitment never came, and the fair settled for saluting the great nation to the north and good neighbors in general. The theme was also chosen as one of a series with international overtones that might complement the growth of locally-based Braniff Airlines which recently had expanded its routes to include several destinations in Europe.

Canada presented a three-fold display which used multi-media and audience participation techniques to depict the country's scenic majesty, architectural heritage and energy development programs.

Another major exhibit was produced by United States government agencies to provide service information. Automotive safety, energy conservation, the population explosion, counterfeiting and credit opportunities were some of the topics covered.

Pepsi Cola sponsored a skateboard team exhibition, and visitors discovered a delightful event in the livestock area — a National Miniature Horse Show for perfectly proportioned mares and stallions standing less than 34″ tall.

Members of the Pepsi Cola Skate Team demonstrated the fine art of skateboarding in 1979.

A Cowboy Heritage Festival, cooperatively presented by the Dallas Historical Society and Austin College, featured storytelling, campfire songs, photos, films and a symposium on western values with such noted panelists as author Larry McMurtry and artist Tom Lea.

The Tony award-winning hit "Annie," based on the legendary comic strip character, captivated fairtime audiences. The top ticket price had risen to $15 and many performances sold out completely. The Music Hall experienced a phenomenon previously reserved for football games — ticket scalping on the front steps.

It was obvious that "Annie" would establish a new all-time top gross, and there were indications that nearly every income area would exceed previous highs. By 4 p.m. on the final Sunday, State Fair officials were waiting for tabulations, expecting and receiving statistical confirmation that the 1979 fair was indeed the most successful and problem-free event in the organization's history.

Word reached staff members in whispers or terse warnings: "The Skyride is down." In midway parlance, that statement meant either that the ride was not operating, had been shut down for some reason, or, as was the case on this sunny October afternoon, there had been an accident.

A fireman helps passengers out of one of the gondolas left stranded on the aerial cable following the Swiss Skyride accident in 1979.

In a sequence of events, as reconstructed by witnesses, one of the gondolas on the mile-long aerial tramway inexplicably stopped near a support tower. It was struck by a second car, then a third, and when another gondola slammed into the three stalled cars, the middle two were jarred loose from the cable and fell 80′ into the crowded midway. One car plummeted through the awning over a game concession and crushed a fairgoer. The other caught on the canvas covering an adjacent booth.

The ride was halted, leaving about 85 people suspended in cars until rescuers using cherry pickers and aerial ladders could bring them down. General manager Gallagher closed the fair and cleared the grounds. Sometime after 8 p.m., the last of the stranded passengers was safely evacuated.

One man was dead; approximately 15 others were injured, one severely. The fair was over; the questions, investigations and legal proceedings were just beginning.

1980

By definition, an accident is a happening that is not expected, foreseen or intended. Critics, and the State Fair had many in the emotional aftermath of the Skyride tragedy, accused the fair of negligence or callous

Food and fairtime are synonymous in the minds of many from the prize-winning jams and jellies (top) to the irresistible sweet treats (bottom) offered at 200 concession stands and restaurants on the grounds.

disregard of safety precautions. Some of the individuals or groups making public statements during this period were motivated, at least in part, by reasons of self-interest, but there was also a widespread sense of anger, a feeling of betrayal by an organization that people loved and trusted.

The fair, on the other hand, rightfully objected to being characterized as an uncaring, make-a-buck operation. Safety had long been a priority concern, and ride inspections that went beyond anything required by law had been carried out at the fair's expense. Obviously, the system had failed to prevent what it was intended to prevent, but the fair, stunned by the accident and unaccustomed to dealing with hostility, reacted defensively. Efforts to determine what had happened and why were further complicated by an adversarial relationship that developed between the fair and the Consumer Product Safety Commission. The issues in question related to legal jurisdiction and the impact on insurance coverage if unauthorized inspections were permitted, but the resulting court action was generally perceived as an attempt by the fair to duck a federal inquiry because it had something to hide. By the time a ruling was handed down, the fair had won a legal battle but lost a war of public opinion.

The investigative process dragged on for months, and years would be required to unravel the complexities of the accident and reach settlements.

By the beginning of the new decade, the population of the United States had reached 226 million. There were 14 million Texans, and 904,078 lived inside the Dallas city limits. In 1980, President Carter struggled with the lingering Iranian hostage crisis, the decision to boycott the Moscow Olympic Games, and a formidable challenge to his second term hopes mounted by the 69-year-old governor of California, Ronald Reagan.

Dallas residents suffered through the longest, hottest summer in the city's history — 52 consecutive days with temperatures above 100 degrees. Fair Park was the subject of another consulting study, and work began on a 2.1 mile segment of State Highway 352 running parallel to the park's southwestern fence line. When completed, the four-lane divided highway would permit an expansion of the park perimeter to encompass an 18-acre area for parking near the Music Hall and museums.

The fair was in a position to reinvest $1 million of its 1979 income in structural improvements including modernization of the Cotton Bowl press box and construction of a new loading station for the Comet Coaster. Three staff members were promoted to the position of assistant general manager with specific areas of responsibility: Bob Halford in marketing and public relations, Claude Perry in administration and finance and Dick Potticary in plant engineering.

Plans for "Around the World in '80" focused on an appearance by the Royal Canadian Mounted Police Musical Ride as the rodeo's featured entertainment and "Camelot" in the Music Hall with Richard Burton recreating his original Broadway role.

The attraction lineup also promised one of the brightest new groups in country music, Alabama, playing for a Saturday night street dance; "Healthworks," a giant exhibit offering free medical screening; recreation of old time radio shows every night by the Texas Broadcast Museum; "Hands Around the World," a collection of paintings by a multi-national organization of children; and two Australian folk music groups playing lively bush songs and pub ballads.

Inflation nudged the adult gate admission up another 50 cents to $3.00. With intensified vigilance in safety areas and with full awareness that people would be watching for any misstep, the State Fair opened on a positive note with Texas' first Republican governor since Reconstruction, Bill Clements of Dallas, cutting the ribbon.

At the end of 17 days, the 1980 exposition would be accorded a ranking in the all-time top five revenue producers. "Camelot" would smash all records by earning a $1.1 million gross. Rodeo attendance would rebound 22%. Yet 1980 would be remembered by fair management as a year in which whatever could go wrong did, without good reasons, at inopportune times, in the worst possible places, and every incident was magnified in a relentless media spotlight.

One way of looking at the State Fair of Texas would be to compare it to a city of 175,000, squeezed into a 277-acre area, with a street life characterized by crowds, physical exertion, frequent money transactions, variable weather elements, and constant interaction between people and high-tech machinery specifically designed to produce a realistic illusion of danger. Under these conditions, statistical probability predicts a certain number of injuries, crimes and mechanical

Garland Parnell and his simian friends are one of the fair's strolling entertainment acts.

Auction sales are a key to the success of the Pan American Livestock Exposition. By 1980, approximately 12,000 international guests had visited the show and purchased more than $74 million of purebred livestock for import to other countries.

malfunctions during any two week period. But statisticians deal in averages, and 1980 was not an average year at the State Fair of Texas. Problems with rides, though none with serious consequences, were more frequent than usual. There were two heart attack deaths prompting questions about emergency medical aid on the grounds. On a purely numerical basis, overall crime at the fair had been reduced, but violent crimes, in particular incidents of robbery and rape just outside the park, increased. And even attendance slipped below the three million mark for the first time in ten years.

1981

The latest in a long line of Fair Park development studies was released in the early part of 1981, and while loaded with graphs, charts, tables and interesting ideas, the report seemed to raise more questions than it answered. The Park Board and the State Fair together with the boards of the museums agreed and disagreed about specific proposals, but one immediate outcome was to facilitate a closer working relationship among the organizations involved in the park's operation. The study also provided a base for bond funding recommendations which would affect Fair Park.

In preparation for the 1981 event, "Going Places, Doing Things," the State Fair upgraded its medical program by purchasing mobile electric ambulances, small cart-like vehicles capable of moving through crowds, staffed with paramedics trained by the Dallas Fire Department. Increased emphasis on security produced towering new light standards in the parking lots, additional police personnel assigned to the four blocks surrounding the park perimeter and the use of officers on horseback to patrol remote areas.

Just prior to the fair, work was completed on the extension of State Highway 352, and the roadway was named Robert B. Cullum Boulevard. Bob Cullum, hospitalized with a circulatory ailment, talked his doctors into a brief leave of absence, took a ride on his namesake thoroughfare and was honored at a dedication ceremony.

The fair had announced "Barnum," starring Stacy Keach, as its annual Music Hall production, but "Barnum" encountered problems on the early leg of its national tour and closed. A revival of the 75-year-old George M. Cohan show, "Little Johnny Jones," with one-time teen idol David Cassidy in the lead was booked as a last minute replacement. Though eminently likeable, Cassidy was no Cohan or Cagney in terms of talent. Fair audiences stayed away in droves, and Donny Osmond took over the title role after the Dallas run.

Street dances are scheduled on Saturday nights, but impromptu street dancing occurs anytime a good country/western band gets warmed up.

The World Congress Santa Gertrudis Show, incorporating important seminars and sales, highlighted the Pan American Livestock Exposition. More than 2,600 cattle were exhibited. A competition for Peruvian Paso breeders was added to the usual lineup of horse shows.

Two multimedia presentations of Texas history were featured in 1981. Concessionaire Allen Weiss opened a new Old Mill Inn featuring favorite Texas recipes. Pepsi Cola sponsored a first-time outdoor ice show on a 20′ x 30′ rink built in the middle of the Esplanade. And visitors welcomed the return of Borden's Elsie complemented with an appearance by a less docile, but equally well-known trademark bovine, the Schlitz Malt Liquor Bull.

The good news about the 1981 fair was that there were no ride-related problems, minimal medical incidents and crime statistics dropped 10%. The bad news was that it rained on nine of 17 days with a discouraging effect on attendance and income figures. Fortunately, a large share of the revenue was

Ron Urban brings his "Pepsi on Ice" show from Canada for an outdoor audience seated on the banks of the Esplanade.

recovered the weekend following the fair when the Cotton Bowl hosted two successive, 80,000-ticket-sellout concerts on the Rolling Stones Farewell Tour.

1982

"Whenever a person was in need, Bob Cullum was there," said Dallas mayor Jack Evans on the death of the longtime community leader in early December 1981. Though the statement applied equally to the needs of the city, the State Fair and countless civic organizations, the loss of Bob Cullum was most appropriately expressed, as it was felt, in personal terms.

State Fair directors Ebby Halliday and John Stemmons coordinated a fund-raising effort to plant the median strip on Cullum Boulevard with 242 live oaks and 800 crepe myrtles as a memorial. A $150,000 beautification program, financed entirely by donations from private citizens, got underway the following March.

For its major 1982 construction project, the State Fair renovated the old Gas Building to serve as a Fair Park information center. A first aid station, new restrooms and offices for the Sesquicentennial Committee's management staff were built into the remodeled facility.

In response to increasing interest in physical fitness and good health habits, the State Fair produced a consumer-oriented show over the July Fourth weekend. The first Fitness Fair offered three days of exhibits, clinics and demonstrations, but failed to attract much of an audience.

Individuals and organizations rallied behind a 1982 bond proposal for Fair Park improvements.

Passage of an $18 million proposal for Fair Park improvements, as part of a $247 million city bond program, received the highest priority on the fair's off-season agenda. The park's supporters and suppliers banded together as Friends of Fair Park and got involved in literature distribution, phone banks and rallies. The proposal, which encompassed badly needed sidewalk, street, waterway, lighting and building improvements plus expanded green spaces, won 3 to 1 voter approval.

"Texas at Its Most" was picked as the 1982 theme. Special features for the annual fall extravaganza included Yul Brynner's final tour in "The King and I"; an environmental project, "Inherit the Earth," produced jointly by the five Fair Park museums; two sensational new amusement rides, the UFO and Looping Star; and a State Fair visit by King Olaf V as the highlight of Norwegian Day.

Big Tex wore a new shirt cut from 100 yards of a burnt orange colored fabric, suggesting to some a partisanship toward the University of Texas football team. The heavily favored Longhorns needed more than Tex's support, as the Oklahoma Sooners triumphed 28-22.

Norway's King Olav, assisted by Governor Clements, clipped the ribbon signaling the fair's official opening. The king made a brief speech and good-naturedly endured unsuccessful efforts to find a cowboy hat to fit the royal head.

Fairgoers, always alert to the unusual, stopped to watch safety skits presented by Mattel's new Monchhichi Monkeys. The giant of the toy industry had become a State Fair exhibitor the previous year with an outdoor display introducing Barbie's Pony to families ripe for a little pre-Christmas promotion.

Foam alligators, complete with wire leashes, topped the novelty sales charts. The three-foot long reptiles followed "topper boppers," water-filled whistles, Chinese yo-yos, "invisible dogs" and the all-time champion fads, live chicks and chameleons — now banned by the SPCA.

The exposition's most spectacular unplanned moment occurred on a Monday night. Following the evening parade, the driver of a stagecoach failed to return his unit to its assigned tent per instructions, instead taking it to the front of the Band Shell to attract attention for an 8 o'clock music show. Curious fairgoers climbed on board the parked coach. At 7:50 p.m., the nightly fireworks began with a barrage of skyrockets and flares launched from a restricted area — directly behind the Band Shell. Frightened by the noise, the team of six Belgian horses took off at high speed and stampeded through the crowd along First Avenue. Three alert motorcycle patrolmen gave chase with blaring sirens and flashing lights and eventually halted the runaways, preventing the incident from causing serious injuries.

Attendance was good at the 1982 State Fair of Texas, but revenue was spectacular, permitting the organization to invest an ambitious $1.9 million in park improvements as part of continuing preparations for a Sesquicentennial celebration in 1986.

The fair serves as a show window for such high ticket items as fur coats, boats, organs and spas (top) and a test market for the newest novelty products (bottom).

1983

Close observers viewed the 1982 fair as perhaps the finest in the organization's long history. The attendance figure of 2,868,062, though not as large as some reported in the 1970s, probably reflected a more accurate tabulation of those who passed through the gates. But fair officials were no longer so enchanted with numbers for numbers' sake. The park was already filled to capacity on

Joe M. Dealey, State Fair President, 1983 —

weekends and Friday and Monday school days. Changing work and social patterns made it unlikely that middle of the week attendance could be increased significantly, leaving these days pleasantly light for the enjoyment of senior citizens, handicapped children and other special groups. Criteria for determining an outstanding year had shifted from attendance to such factors as program quality, operation efficiency, satisfactory performance in critical safety areas and public response as gauged by participation, enthusiasm and spending. On this scale, 1982 appeared to be a strong contender for "best ever" honors.

It was not a time for resting on laurels, however. To complete the construction agenda, it was necessary to begin at once on a list that included paving the area adjacent to Cullum Boulevard to provide 1,200 parking spaces, new gates for the Grand and Martin Luther King entrances, a street from the midway arch to the roller coaster to improve pedestrian traffic flow in the amusement area, a new $300,000 log flume ride, additional midway restrooms, roofing for the Automobile Building and a $250,000 renovation of one end of the Embarcadero to create a new home for the creative arts department. This last project involved all-new display cases, two kitchen theaters for demonstrations or competition and a performance stage with permanent seating.

The fair's directors voted to enlarge the board from 50 to 60 to permit the addition of members from other parts of the state, and they elected Joe M. Dealey, publisher of the *Dallas Morning News,* to replace the retiring Al Davies as president.

Seldom are there vacant seats for any performance of the State Fair-Gil Gray Circus presented each year by the Dr Pepper Company.

Cotton Bowl concerts, other stadium events such as motorcross and tractor pulls, popular flea markets, antique shows and ethnic festivals, plus museum and Music Hall activities, attracted large crowds to the park, but the State Fair's second annual Fitness Fair, moved to an earlier date, flopped as badly as the first, leading management to conclude that people interested in fitness preferred to spend a spring weekend jogging, cycling or boating rather than visiting a show that explained jogging, cycling and boating.

"Lena Horne: The Lady and Her Music," the dazzling one-woman show originally set to open the remodeled Majestic Theater in downtown Dallas, was signed as the State Fair Musical. Miss Horne had cancelled the Majestic booking after a local political squabble developed over ticket prices and minority interests.

The 1983 fair was billed as "Best of Shows" and boasted an increased number of attractions which were free to visitors once inside the gates. The key to improved entertainment at no extra cost to the fair was a growing program of corporate sponsorship. The fair had enjoyed these mutually beneficial relationships in the past, but by 1983, marketing wizard Bob Halford had put together attractive packages with high visibility that could be matched to the aims and budgets of various businesses. The soft drink companies — Dr Pepper, Coca Cola and Pepsi Cola — continued to support respectively the circus, stage and ice show that bore their names. Three beer companies — Schlitz, Pabst and Stroh — sponsored outdoor musical entertainment. Budweiser paid for the new Cotton Bowl scoreboards; Coors financed a nightly laser show; and Miller assumed responsibility for the Rock & Roll Time Machine, a multimedia adventure in a giant geodesic dome. Additionally, Trailways took on some of the costs of the in-grounds transportation system, and Texas Instruments supplied equipment and know-how for computerized information centers. This infusion of outside money allowed the fair to reallocate its budget to build and book indoor stages and hire strolling entertainers to mingle with the crowds.

For the first 10 days, the 1983 fair appeared to coast on the momentum created the previous year, but on the second Monday, shortly after 8 p.m. nearing the end of a busy school day, a car tore loose from the spinning Enterprise ride, hit the ground and slid about 40′ out into the congested midway. One of the teenage passengers in the car was killed, the other two seriously injured.

Most people learned about the accident during the 10 o'clock newcasts. Viewers were given summaries of the available facts combined with eyewitness accounts, then many watched an interview in which a bystander who represented himself as a ride operator told a bizarre tale of inside knowledge about unsafe machinery and total absence of

Corporate sponsorships have permitted improvements in entertainment and visitor services at no additional cost to the fairgoer.

Fiddlin' Frenchy Burke, a dynamic performer on the Coca Cola-Big Tex Stage.

inspections, after which he dramatically announced he would resign his position with the State Fair. Though investigation later revealed that the individual was not a ride operator and had never been employed by the fair, his fabricated statements were picked up and carried on national wire services.

But inaccuracies in reporting did not obscure the real concerns of the public and of fair management — a second accident involving a fatality had occurred within a four year period. The 1983 exposition finished the remaining six days of its run, and the investigative process began again.

1984

Americans welcomed 1984 with an invigorated economy and renewed national spirit. The country mirrored a conservative, business-oriented outlook, and Dallas aspired to establish itself as a role-model "City of the 80's" when it hosted the national GOP Convention in August.

Other cities west of the Mississippi nurtured ambitious hopes for 1984. The Democrats had selected San Francisco for their convention site; the Summer Olympic Games were scheduled for Los Angeles; and New Orleans would have its opportunity to present a world's fair.

Two Texas businessmen, pursuing a dream to add Dallas to the elite list of international cities on the Formula One Grand Prix Circuit, had secured necessary approval and funding to build a 2.43 mile course in Fair Park for a July race date. To accomplish this in less than four months, earth moving and paving equipment rolled into the grounds and rearranged the landscape. At the same time, the Park Board was firming up plans for spending bond money on a redevelopment program,

Starship 3, a helicopter aerobatic act, was one of the fair's new attractions in 1984.

The good guys and the bad guys tangled in another first-time fair event — Cotton Bowl Wrestling.

which guaranteed that the park would be torn up again through most of 1985 and 1986.

The State Fair was coming off a good year financially and starting to lay groundwork for a major celebration of the Texas Sesquicentennial in 1986. One of the first steps was to schedule a 24-day fair in 1984 to see what impact a longer run might have on exhibitors, concessionaires and the audience.

But with all the excitement and projects in the wings, fair general manager Wayne Gallagher had dedicated himself first to putting together the most comprehensive ride

Visitor surveys are a reliable, though certainly not infallible, source of information in the outdoor amusement business. A 1984 study by Economic Research Associates of Los Angeles offered the following analysis of the State Fair of Texas:

A majority of visitors (78%) live in an area of North Texas defined by Dallas, Tarrant, Collin, Denton and Rockwall counties.

Most people come to the fair with other family members and stay on the grounds for about six hours. They arrive in groups of three or four persons, and each individual spends about $7.25.

Among adults surveyed, the average fairgoer is under 40 (64%), married (72%) and has at least a high school education (90%). More women attend than men. Many will visit the exposition more than once, and nearly 50% have seen each of the past five fairs.

Visitors rate commercial exhibits, food, livestock activities and free entertainment, in that order, as their favorite parts of the fair.

The grounds are four times more crowded on weekends than mid-week days. On a big Saturday or Sunday, when attendance exceeds 250,000, 60% of these fairgoers will be in the park at the same time during the early afternoon.

The State Fair of Texas has 76 staff members on its year-round payroll and hires 1,782 temporaries for the extended fairtime period. Another 16,370 workers are employed by fair-related enterprises. Out-of-town performers and participants, with living expense requirements during all or part of the run, number 8,680.

The total impact of the State Fair of Texas on the Dallas economy is calculated at $127 million.

inspection/ride safety program ever attempted by a fair. With solid backing from President Joe Dealey and the board of directors, Gallagher spent months conferring with knowledgable sources and developed a set of procedural guidelines which earned praise from the amusement industry and even won endorsement from the once-hostile Consumer Product Safety Commission.

The basic elements of the program called for year-round records kept by ride owners to show compliance with all manufacturers' maintenance and safety bulletins plus a log that explained any repairs; detailed, stripped-down inspections of the machinery by a team of experts hired by the State Fair and sent to off-season locations around the United States to check rides that had applied for Dallas bookings; another thorough pre-opening inspection with operational testing; an inspector to remain on site throughout the run to make intensive equipment checks three times weekly; and a midway guest relations booth to handle fairgoers' comments and complaints.

"We can't guarantee perfection," said Gallagher responding to queries just prior to the fair. "But we are trying to do everything we can and go far beyond what anybody else has done . . . We'll know Friday."

"OctoberBest," the 1984 theme, described an extended event packed with new features including the burlesque musical "Sugar Babies," with Mickey Rooney and Ann Miller; a helicopter aerobatic act performed 100′ above the lagoon; an indoor truck and tractor pull; judging of Simbrah cattle; four model homes furnished with modular furniture; a Delta Airlines 727 jet engine on public display; plus Cotton Bowl Wrestling, a Saturday afternoon marathon of matches spotlighting the enormously popular Von Erich brothers.

The Texas Department of Agriculture presented a special exhibit: "Texas Crossroads: the Folk Arts of Agriculture," which offered artifacts from Texas farms, ranches and fishing operations, as well as daily demonstrations of such crafts and skills as saddle making, quilting and blacksmithing.

Management felt confident going into the 24-day run. Changes in statewide policies governing school attendance had derailed plans to expand the fair's school day program, and there were indications that these stricter regulations might impact on the junior livestock program and high school band competition, but generally speaking, all signs pointed to an outstanding exposition. The economy was healthy; the entertainment lineup was exceptionally strong; response to the safety program was positive. All the show needed was its fair share of sunshine.

What it got was an un-fair, unbelievable amount of rain — 17 out of 24 days. In one of the soggiest expositions ever, a

One of the famed Budweiser Clydesdales stands next to a miniature horse while both wait to appear on a noon newscast being televised live from the fair.

surprising 2,931,691 visitors braved the elements to see the sights. The most common sight was clusters of wet fairgoers huddled under building eaves waiting for a break in the clouds.

The midway was forced to shut down repeatedly. Parades and fireworks were cancelled. Entertainment schedules were reshuffled. Outdoor concessionaires and exhibitors suffered. Medical personnel reported no serious injuries, but an increase in strains, sprains and abrasions for people who slipped on wet walkways.

"It's not a disaster, but it sure is a disappointment," said one staff member summing up management's reaction. The silver lining inside the clouds was the faultless performance of the new ride inspection program and continued improvement in all areas involving public safety. The fair's biggest concern was the effect this might have on plans for the Sesquicentennial. The weather had obscured any valid analysis of an extended run, and though the organization finished the year in the black financially, profits from 1984 would not cover as many 1986 projects as once anticipated.

For what was certainly not the first nor likely the last time in its 100-year history, the State Fair of Texas went back to the proverbial drawing board to reconsider, revise, rebudget and ready itself for another go-around and another chance at the brass ring.

And Much, Much More!

1985-1986

1985

Though the State Fair of Texas has never failed to open on time, no one connected with the organization would suggest that it runs like clockwork. The fair has a century-long record of starts and stops, peaks and gullies, bold advances, strategic retreats and frequent turn-arounds. Reflecting the nature of its product, the organization has survived close calls, near misses, even premature burials by what modern idiom might term "hanging in and hanging loose." The fair has proven itself an amazingly resilient operation, one with enormous appetites and ambitions and one capable of taking spectacular pratfalls in full view.

Each fair bears some resemblance to art by committee. Internally, the organization consists of a dozen or more separate "kingdoms," each vying for space and attention, each responsible to an independent set of constituents. A budget is drawn up, but there are always variances. A master calendar is created, but it is often amended before the ink can dry. Rules exist — with numerous exceptions.

Construction on the city's $18 million renovation of Fair Park began in February 1985.

The only thing more complicated and less orderly than putting together a fair would be putting two different fairs together simultaneously, which was the challenge facing the organization at the begining of 1985.

A major exposition, such as contemplated to celebrate the Texas Sesquicentennial, obviously required extra lead time to secure quality exhibits and entertainment which could sustain an extended run. By the same token, a blockbuster 1985 fair was vitally important to generate revenue to carry out the 1986 program. To this dilemma was added an element of suspense and confusion, as construction began on the city's 18-month, $18 million facelift for Fair Park, a project that would of necessity be interrupted in order to have a 1985 fair.

The objectives of the redevelopment program placed emphasis on safety improvements, visual impact, optimum year-round return, complementary immediate and long range planning, compatibility with existing architectural design and preservation of historically significant facilities within the park. Bond money, when combined with contributions by the fair over a five year period, provided a $25 million nest egg for accomplishing these goals.

The partnership between the State Fair of Texas and the City of Dallas Park Board has been described as a successful marriage. The analogy is appropriate. Any marriage that survives more than 80 years, strengthens both parties and from this strength creates a bountiful estate for the beneficiaries of the union, certainly can be termed successful.

But to see the relationship as "loving and harmonious," on a day-by-day, year-in, year-out basis, would require rose-colored glasses, or on some occasions, an inch-thick blindfold.

The first disagreement occurred before the honeymoon was over. Fair president C. A. Keating signed the original contract under protest and only because it was time to open the 1904 event. He warned he would be back to have provisions rewritten. He was, and some were.

Since that time, mixed in with all the joint ventures and cooperative efforts, there have been differences of opinon, delays, bullying by strong-willed leaders on both sides, jockeying for control, hard-feelings, hard-headed foot-dragging — in sum, about what might be expected in a partnership of this duration involving some of the most powerful men in the city's history. And one woman.

Betty Marcus became president of the Park Board in 1982 when Starke Taylor resigned that office to run for mayor. Dallas voters had overwhelmingly endorsed an $18 million bond proposal for Fair Park improvements. Under Mrs. Marcus' direction, plans were drawn up for a far-reaching redevelopment program. Her untimely death in 1983 left completion of the project to the capable leadership of her successor, Billy Allen, the first black to serve as Park Board president.

An army of 80 construction workers supplemented by a motorized division of bulldozers and Mack trucks invaded the grounds in February. Within days, Fair Park looked like a war zone. Mountains of rubble, chunks of concrete and twisted steel reinforcements covered the landscape. Trees were uprooted and replanted on the back acres as the demolition forces began at the main gate and moved relentlessly along both sides of the Esplanade across Grand Avenue and encircled the museums and lagoon. Water, gas, telephone and electric company workers followed laying trenches for new utility services. When completed, this area would be transformed into a pedestrian park with brushed concrete walkways, grassy plots, fountains, deco-style benches, planters and light fixtures plus flower beds, shrubs and trees. Perimeter fencing improvements and installation of twenty-one 100′ light towers were incorporated into the master plan as safety features. The Automobile Building was scheduled to be gutted and rebuilt with additional restrooms and a new heating and air conditioning system. Plans specified a remodeled facade that would recapture the artistry of the original building destroyed by fire in 1942. The Band Shell was marked for plumbing and electrical renovation, new seating, planters and paint.

"Something New for You" was chosen as 1985's fair theme, a promise of glories to come and an explanation for the less-than-finished look which would greet fairgoers in the fall.

Community interest in the park's future was evidenced as a non-profit support group formed and assumed the name: Friends of Fair Park. This organization's goals included funding for projects such as the restoration of the art deco murals on the exterior of the Centennial Building and renovation of the old Margo Jones Theater.

The Broadway hit "42nd Street" is based on the glittery movie musicals of the 1930s.

But of all the projected changes in the appearance of Fair Park, clearly the most visible new element would be the giant ferris wheel, 212′ tall, the equivalent of a 20-story building, to be built on the midway by concessionaire Buster Brown. Capable of carrying six passengers in each of its 44 gondolas, the Texas Star, outlined by 15,000 red, white and blue bulbs, promised to enliven the Dallas skyline with a spectacular computerized light show.

A high-dive aquatic show and the first southwestern tour date for Broadway's song and dance fable "42nd Street" were among the early entries on the 1985 program. By summer, exhibit and concession contracts were in the mail; outdoor entertainment was booked; and sponsorships were under consideration.

With the 1985 event taking shape, fair management sat down to hammer out a format for 1986. After tinkering with the number of days, the operating hours, the possibility of split

The theme poster for the 1985 State Fair.

schedules, selective openings and building change-outs, the finalized plan addressed the problems presented by short-term shows and high-traffic, single-performance special events. It also confronted the reality of basic operating costs which approached $225,000 per day regardless of how many or how few visitors passed through the turnstiles. Unlike the 1936 Texas Centennial Exposition, there would be no state or federal funding, nor would there be a temporary organization created to bear the financial and managerial responsibilities. The 1986 event would be a State Fair of Texas production with all the attendant risks and rewards. And unlike the commissions charged with presenting the recent world's fairs in New Orleans, Knoxville and Montreal, the State Fair of Texas planned to be in business again in 1987.

In June of 1985, the fair announced a 58-day event to honor the founding of the Republic of Texas. "Texas 150" would run from August 30 to October 26, 1986 as a two-fold celebration. The first 27 days would be highlighted by weekend shows, major concerts, ethnic festivals and the opening of a major pavilion, dedicated to the reenactment of Texas history. During this period the public would be admitted to the park and special exhibits without charge. A festive downtown parade and nostalgic renewal of the SMU-TCU football rivalry in the Cotton Bowl would kick off the 31-day, exposition-configuration phase. Additional pavilions scheduled to open would focus on China, the NASA space program, robotics, petroleum, education, travel and communications. National livestock shows and statewide creative arts competition would be featured along with standout entertainment attractions such as the smash London and Broadway musical "Cats."

Skeptics and traditionalists frowned because "Texas 150" didn't fit a familiar pattern. Those who wanted to relive 1936 were disappointed. Those who recognized the enormous changes in society and entertainment standards since that hallowed exposition hailed the concept as innovative and intriguing.

Counting on bandwagon support from the community, the State Fair pressed forward, juggling arrangements for the upcoming fair and the not-so-far-off celebration, hoping that nature would balance the books and bless 1985 with sunshine. It was a show of organizational bravado and fire-tested confidence — a "keeping the faith" pact with Billy Gaston, Sydney Smith, Alex Sanger, Ed Kiest, Bob Thornton, Bob Cullum and the people of Texas.

Visitors to Texas 150 will see these priceless terra-cotta warriors and other art treasures in a major pavilion sponsored by the government of China.

It was a little after six, and though the sun was setting, shards of light knifed through the low hanging clouds and rippled over the wet pavement.

Little Joe y La Familia, a young band with Tex-Mex roots and commercial savvy, finished "Jalisco" with a strumming crescendo.

There was a smattering of applause from the 40 or 50 people gathered in front of the stage on Cotton Bowl Plaza. The crowd should have been larger. It was Texas-Mexico Day at the State Fair, and Little Joe had a loyal following. But intermittent rain had fallen throughout the day, as it had the day before, and as it had for more than a week. Though the drizzle let up by late afternoon, the public had grown cautious. So Little Joe was playing to a mixed audience — employees on break, concessionaires, a few hardy fairgoers wearing waterproof jackets and carrying umbrellas, a group of Mexican-American teenagers and one or two families with small children.

Opposite the stage and behind a row of food stands, the midway neon glimmered in the twilight, and the giant rides roared, thumped and hissed over a background of canned music. The Ferris Wheel turned, the Pirate Ship swung its 180-degree arc, with or without passengers.

Joe Hernandez nodded to guitarist Bob Gallarza. The band launched a spirited version of "Proud Mary," and the crowd came alive. In the front row, 56-year old Mattie Jones, who was waiting for her husband — Crazy Ray of Dallas Cowboys' fame — to finish his last show, began to sway with the music. Encouraged by others, Mattie took a few steps out onto the asphalt strip that served as a dance floor. A grinning stagehand — lanky, blond, in his early twenties — grabbed her hand, and the two bounced to the beat in a dance that spanned generations and color lines. A tall, white-haired gentleman joined them, clapping and rocking along on his own, looking for a partner, then finding one — a shapely black girl wearing the bright yellow shirt of the State Fair clean-up crew. She leaned her push broom against a tree and matched steps with the older man, a flash of Tina Turner here, a jitterbug twirl in response.

Young and old. Black and white. A Mexican band playing rock and roll. Sun after rain.

The great State Fair of Texas . . .

INDEX

A

B

C

D

G

H

I

N

O

P

Q

R

S

T

U

V

W

XY

Z

SOURCES

Microfilm copies of the *Dallas Morning News* and the *Dallas Times Herald* provided a foundation for the year-by-year chronicle of the State Fair from 1886 to the present. Daily newspaper accounts were supplemented by photographs, programs, brochures, premium lists and correspondence from the collections of the State Fair of Texas, the Dallas Historical Society and the Texas/Dallas History and Archives Division of the Dallas Public Library.

The corporate history of the State Fair of Texas was based primarily upon an examination of the organization's annual financial reports and recorded minutes from meetings of the State Fair Board of Directors, executive committee and stockholders covering the period from 1905-1985. Additional resources in this area included Sydney Smith's unpublished notes on the fairs of 1886-1910; J.T. Trezevant's history of the early fairs (1886-1904); Leah Jarrett's unpublished manuscript covering the development of the women's department (1946-1964); and interviews with State Fair of Texas directors and staff members whose involvement with the organization spans the past 35 years.

Valuable sources of information on specific topics relating to the fair included Ralph Widener's *William Henry Gaston: A Builder of Dallas* (Dallas: Historical Publishing Company, 1977); William Sewell's master's thesis, "Early History of Dallas Fairs" (Southern Methodist University, 1953); "Centennial History of the Dallas, Texas Park System — 1876-1976," a project prepared for Texas Tech University in 1976 by Harry Jebsen, Jr., Robert M. Newton and Patricia R. Hogan; Wallace Chariton's *Texas Centennial: The Parade of an Empire* (Plano, TX: Wallace O. Chariton, 1979); "Robert Cullum: An Oral History Interview," conducted by Alan Mason on April 9, 1981, and published by East Texas State University and the Dallas Public Library, in 1983; and Frederick Drimmer's *Very Special People* (New York Amjon Publishers, Inc., 1973).

Three excellent books were used to construct a framework of Dallas history: A.C. Greene's *Dallas: The Deciding Years* (Austin: Encino Press, 1973); Darwin Payne's *Dallas: An Illustrated History* (Woodland Hills, CA: Windsor Publications, 1982); and William L. McDonald's *Dallas Rediscovered* (Dallas: Dallas Historical Society, 1978). Other helpful works on Dallas included the following: Sam Acheson, *Dallas Yesterday* (Dallas: SMU Press, 1977); Lucille Boykin and Mary Brinkerhoff, *Tent to Towers: A Dallas Time Line* (Dallas: Dallas Public Library, 1982); A.C. Greene, *Dallas USA* (Austin: Texas Monthly Press, 1984); Philip Lindsley, *A History of Greater Dallas and Vicinity* (Chicago: The Lewis Publishing Company, 1909); and Ted Dealey, *Diaper Days of Dallas* (Nashville: Abingdon Press, 1966). Statistical information was obtained from microfilm copies of Dallas city directories and Texas death records on file in the Texas/Dallas History and Archives Division of the Dallas Public Library.

The dates and facts needed for a broader historical perspective of this era were gleaned from such works as Archie P. McDonald, "Texas: Its First 150 Years," published as a special supplement to the *Dallas Times Herald* in 1982; Michael T. Kingston (ed.), *The Texas Almanac, 1984-1985* (Dallas: The Dallas Morning News, 1984); John A. Garraty, *A Short History of the American Nation* (New York: Harper & Row, 1980); Francis Russell, *The American Heritage History of the Confident Years* (New York: American Heritage Publishing, n.d.); *The Timetables of American History*, edited by Laurence Urdang (New York: Simon

& Schuster, 1981); and the seven-volume series of Time-Life Books, *This Fabulous Century* (New York: Time Inc., 1969).

Two books provided background on expositions and the amusement industry: Kenneth W. Luckhurst, *The Story of Expositions* (London/New York: The Studio Publications, 1951); and Joe McKennon, *American Carnival* (Sarasota, FL: Carnival Publishers, 1971).

Magazine articles used as sources included: Tom Peeler, "State Fair!" *D,* October 1982; The Editors of *D,* "Power in Dallas: Who Holds the Cards?" *D,* October 1974; Mark Goodman, "State Fair: She Crawls on Her Belly Like a Reptile," *Time,* November 1, 1971; David L. Cohn, "Dallas: Capital of the New South," *Atlantic Monthly,* October 1940; "Big Time in Dallas at the Texas State Fair," *Time,* October 13, 1947; and "Dallas in Wonderland," *Fortune,* November 1937.

PHOTO CREDITS

(Key: T=top, C=center, B=bottom, R=right, L=left when more than one photo appear on the same page.)

Collection of the Dallas Historical Society: 1, 2-B, 3, 6-T, 6-B, 12, 15-T, 15-B, 23, 31-T, 61-L, 73-B, 74, 109-B, 116-B, 118-TR, 119-CL, 119-CR, 119-BR, 120-B, 122-TL, 123-CL, 125-TR.

Collection of the Texas/Dallas History and Archives Division of the Dallas Public Library: 2-T, 4, 43, 46, 54-B, 55, 60, 66-L, 83, 86-B, 87, 91, 94, 115, 122-CL, 124-B, 126-TR, 156, 158.

Collection of the Dallas Museum of Art: 118-TL, 118-BL, 123-CR.

Personal collection of Joseph B. Rucker, Jr.: 11-B, 27, 32, 37, 39, 50.

Personal collection of David Nixon: 56-T, 59, 62, 63-T, 64-T, 118-CL, 120-TL, 120-CR, 121-CL, 121-CR, 124-CR, 127, 128-L, 129-T, 129-B.

Personal collection of Charles Kavanaugh: 118-BR, 125-CL.

Personal collection of Wayne H. Gallagher: 68.

The Circus World Museum, Baraboo, Wisconsin: 25-B, 80.

The Bettmann Archive: 34, 35, 47, 69-T.

The Minnesota Historical Society: 88-B.

Andy Hanson: 174-B, 212-BR.

Dr Pepper Company: 97.

Ice Capades, Inc.: 146.

Chuck Stevenson: 209.

20th Century Fox Studios: 178.

Martha Swope: 231-B.

Shirley Oxendine: 215-R.

All other illustrative materials were obtained from the photo archives of the State Fair of Texas and to the author's knowledge represent photographs made while in the employ of the State Fair of Texas or donated to that organization. Because of age or duplication processes, many of these prints do not carry credit lines to identify the original photographer, but included in the collection are photos by the following: Wayne Miller, Neal Lyons, Jack Beers, Squire Haskins, Clint Grant, Max Ewing, Doris Jacoby, Tom Dillard, Kinkaid Photo, Don Cook, Jeanne Deis, Frank Rogers, Dal-Tex Photos, Bradley Photographers, Cliff Hopper, John J. Johnson, Gildersleeve, Clogenson, Bachrach, Bob Goodman, and staff photographers for the *Dallas Times Herald* and the *Dallas Morning News.*

TEXAS STATE FAIR
THIRTEENTH ANNUAL ENTERTAINMENT TEXAS STATE FAIR. ... 1–16.